Population Ecology

Contents

Preface

This text of computer simulations is designed to accompany an undergraduate college course in ecology, population biology, or conservation. The exercises allow the student to explore, by means of computer simulations, the effects of biotic and abiotic variables on population densities through time. They also provide basic modeling skills that can be applied elsewhere, as in term papers dealing with the conservation of endangered species or the control of pest species. Each exercise can be completed within approximately two hours. The text assumes no prior knowledge of MATLAB by either the student or the instructor.

This version of the text uses MATLAB Release 12 or 13 and Windows 98 or later; it is written for use in a laboratory, in which files are stored on a portable disk. I have assumed, here, that the portable disk is a floppy disk. If the disk is not a floppy, then the command `cd a:` needs to be changed, substituting the appropriate letter other than `a`. If the exercises are done on a student-owned computer, then the command `cd a:` should be omitted wherever it appears. Output is in the form of graphs with figure legends, in which the student demonstrates an understanding of the biological principles. The exercises on predation, optimal foraging, conserving an endangered species, and controlling a pest species can be used as class projects, in which the class is divided into groups with each group responsible for one aspect of the topic.

The exercises are presented in as simple a form as possible, in order that the student can concentrate on biological topics rather than on simulation procedures. New techniques and shortcuts are introduced gradually so as not to overwhelm the student with procedures in the first few weeks of the course. Commands are explained each time they are used for the first time and are defined in the glossary. In addition, the software program can be used to define commands. To find an explanation of the term `dsolve`, for example, simply enter : "`help dsolve`."

Three windows are typically used when carrying out the simulations: the command window, the command-history window, and the program window. The command window is the main window, in which most information is entered. It always displays the symbol >> in the left margin. The command-history window keeps a record of all commands that are entered. It can be used to search for errors and to transfer commands (by highlighting, right clicking, copying) to the command window (by pasting). The program window is used to write and store programs (called *m-files*) that will be applied more than once. The command window and the command-history window are opened by scrolling down "view" on the toolbar; the program window is opened by scrolling down "file" and then either "new, m-file" or "open" on the toolbar. Always clear both command windows (scroll down edit) between exercises when doing more than one exercise in a single sitting.

This text has been tested on several classes of undergraduates at the University of Colorado. Thus, if part of an exercise does not work, most likely it is because you have made a typing error. If you cannot find the error, it helps to print your commands and then review them carefully on the printed copy.

I am deeply indebted to Joan Roughgarden for this manual, both for the general concept of teaching population ecology by means of computer simulations and for specific techniques presented in her text, *Primer of Ecological Theory* (Prentice Hall, 1998). The advantage of using this approach to teach the dynamics of populations, as compared with either no simulations or simulations that use "canned" programs, is that students learn the logic of the classic models, how changes in the variables influence population densities, and how to develop their own models of population change through time.

I acknowledge also the exceptional support provided by Rachael Ballard, Assistant Editor at John Wiley & Sons Ltd, who believed in this project from the beginning and provided encouragement and assistance throughout the reviewing and revision process.

I thank Robert Hambrook, Senior Production Editor, and Neville Hankins, Copy Editor, for their highly professional production of the text. Most of all, I thank Stephen Bernstein for assisting in so many ways throughout the entire project.

Ruth Bernstein

Acknowledgements

This book makes extensive use of the MATLAB program, which is distributed by The Mathworks, Inc. We are grateful to The Mathworks for permission to include extracts of this code.

For MATLAB product information, please contact:

The MathWorks, Inc.
3 Apple Hill Drive
Natick, MA, 01760-2098 USA
Tel: 508-647-7000
Fax: 508-647-7101
E-mail: info@mathworks.com
Web: www.mathworks.com

A user with a current MATLAB license can download trial products from the above Web site. Someone without a MATLAB license can fill out a request form on the site, and a sales rep will arrange the trial for them.

EXERCISE 1

Exponential Population Growth

A small population in a favorable environment can grow at a rapid rate. There will be few deaths, except from old age, and the rate of reproduction will far exceed that needed to replace these deaths. This type of growth, known as exponential growth, can be modeled either in continuous time (by differential equations) or in discrete time (by difference equations).

Populations with Continuous Growth

A population that has no defined reproductive season exhibits continuous growth, because births occur at all times of the year. Human population growth is an example of continuous growth. The increase in numbers of individuals through time is described by the differential equation

$$\frac{\mathrm{d}N}{\mathrm{d}t} = rN \tag{1.1}$$

where $\mathrm{d}N$ is the symbol for change in numbers and $\mathrm{d}t$ is the symbol for change in time. Thus, $\mathrm{d}N$ over $\mathrm{d}t$ means the change in numbers that occurs over a particular period of time. The symbol r is the population growth rate per individual (i.e., the rate at which the

Population Ecology: An Introduction to Computer Simulations. By Ruth Bernstein.

population increases in numbers divided by the total number of individuals in the population at that time); it is a constant in the equation. The symbol N is the number of individuals in the population at each point in time. This differential equation, which describes the instantaneous rate of increase, can be used to find an equation that predicts the number of individuals in the population at some future time t, an equation in which the slope at each point in time is equal to r times the number of individuals in the population at that point in time. To have MATLAB find this predictive equation for you, follow the instructions below.

Open Microsoft Word and press *enter* a few times so that your first graph will not be at the very top of the page. Open MATLAB on top of Word. In the MATLAB command window (which has the prompt >>), type the following:

```
>> nt=dsolve('Dn=r*n', 'n(0)=no', 't')
```

The command `dsolve('Dn=r*n')` tells the program to solve the differential equation defined by `r` multiplied by `n` (MATLAB uses a small `n` for N). An asterisk means multiply. (Each equation is always identified by enclosure within single quotation marks.) The command `'n(0)=no'` tells the computer that the initial number of individuals `n(0)` is defined by `no`. The time period during which the population grows is identified by `t`. MATLAB answers with

```
nt=no*exp(r*t)
```

which is the equation for predicting the population size at any time. It is computer language for

$$N_t = N_0 e^{rt} \qquad (1.2)$$

Now use MATLAB to predict the growth of the US population over the next 100 years. The present population (i.e., `no`) consists of about 300 million and the growth rate (`r`) is 0.017 per individual per year. When a computer program is used in calculus, continuous time is approximated by having time change in tiny steps from time 0 to time t. Here, time increments of 1 year give an almost smooth curve. To tell the computer to plot population size over a time period of 100 years, in increments of 1, the command is `t=0:1:100`:

```
>> no=300;
>>r=0.017;
>>t=0:1:100;
```

(A semicolon placed at the end of a command stops the computer from displaying its response to your entry.) To place your results on a figure, simply type

```
>>figure
>>hold on
>>plot(t,eval(vectorize(nt)),'r')
```

The `figure` command tells the computer to set up a graph. The `hold on` command says that you will be placing more than one curve on the graph, so the computer should keep this figure window open until you have added all these curves. The `plot(t, eval(vectorize(nt)))` command tells the program to plot `t` on the X axis and the solution to the equation that predicts `nt` (see above) on the Y axis. The term `vectorize` produces a version of the equation that can be plotted on a graph. The `'r'` part of the command produces a red curve on the graph. (The symbols for other colors and line styles are provided on page 159. If you do not have a color printer, you may wish to use line styles rather than colors to distinguish between the different curves on a graph.)

Population growth is determined by rates of immigration and emigration as well as by births and deaths. In the United States, for example, the actual rate of increase r is 0.017, but without immigration (and the children of immigrants), the population growth rate would be approximately 0.007. To simulate population growth under these conditions, type

```
>> r=0.007;
>>plot(t,eval(vectorize(nt)),'b')
```

Thus, the population would grow even without immigration. The cause of this growth is not a high birth rate but rather a steady increase in lifespan, brought about by medicine and other survivorship-enhancing aspects of the society. When the length of life stabilizes (near the maximum lifespan set by the genes of each individual), the population will decline because the number of children born per couple is less than two: on average, each woman in the United States leaves 1.8 children in her lifetime. If all children survive to reproductive age, then the population growth rate (with neither immigration nor a continuing increase in lifespan) would be: $r = -0.00513$. Simulate this rate of growth by typing

```
>>r=-0.00513;
>>plot (t,eval(vectorize(nt)),'g')
```

In the figure window, label the graph by scrolling down "insert." Then click each label that you want to apply. Give the graph a title and label the axes: the X axis is years and the Y axis is number of individuals (times 1 million). Label each curve by clicking "A" and then clicking a spot near the curve where you type the label. The red curve is with immigration; the blue curve is without immigration; the green curve is with neither immigration nor increasing lifespan.

Move this graph to a Word document by scrolling down the "edit" icon at the top of the graph window and clicking "copy figure." Then open the Word document (by clicking on the window beneath MATLAB or by clicking on the Word icon at the bottom of the screen) and click "paste." Write a legend for the figure. Move the cursor below the figure to position the document for your next graph. Return to MATLAB by clicking on its window. In the command window, scroll down "edit" and click "clear command window."

Populations with Discrete Growth

Now consider ground squirrels, which breed only in the spring. Population growth occurs in discrete time periods rather than in continuous time throughout the year. The population size at time $t + 1$ is given by the difference equation

$$N_{t+1} = RN_t \tag{1.3}$$

where N_{t+1} is the number in the population one year from now, N_t is the number now, and $R = 1 + (B - D)$, in which B is the number of births per individual per year and D is the number of deaths per individual per year. The equation begins with a 1 so that when multiplied by N_t you get N_t, which is the number of individuals in the population before the reproductive season. This R is called the geometric growth factor.

We now want to modify the equation to be able to predict the population size for more than one year in the future. When R is constant, the equation for predicting the number of individuals one year from now can be written

$$N_1 = RN_0$$

and the equation for predicting the number of individuals two years from now is then

$$N_2 = RN_1$$

or

$$N_2 = R(RN_0) \text{ or } N_2 = R^2N_0$$

the equation for predicting the number of individuals three years from now is

$$N_3 = R(N_2) \quad \text{or} \quad R(R^2N_0) \quad \text{which is} \quad R^3N_0$$

and so on for any time period t. Thus, the general equation is

$$N_t = (R)^t N_0 \tag{1.4}$$

which is written in MATLAB language as `nt=(r^t)*no`, where `nt` stands for N_t and `no` stands for N_0, `r` stands for R, and `t` stands for time. The asterisk indicates multiplication. The symbol ^ indicates an exponent; the symbols `r^t` are enclosed within parentheses to indicate that this mathematical procedure (r^t) is to be carried out first – before multiplying by `no`. Enter this equation as

```
>>nt='(r^t)*no';
```

Now assign some values to the parameters. Assume an initial population of 100 ground squirrels. Assume that R (the rate of growth per year) is equal to 1.05 (i.e., each year, the number of individuals in the population is 1.05 times the number of individuals in the population the previous year). Show on a graph how this looks over a time period of 30 years.

```
>>no=100;
>>r=1.05;
>>t=0:1:30;
>>figure
>>plot(t,eval(vectorize(nt)),'b')
```

Label the graph. Move it to a Word document (i.e., in the graph window, scroll down "edit", click "copy figure," and then, in the Word document, click "paste"). Write a figure legend.

EXERCISE 2

Population Invasions

A population in a favorable environment undergoes exponential growth. When such a population is surrounded by favorable habitat, then it will spread outward until it occupies all the available habitat. The appearance of a population in a new area is called an invasion. The rate at which the population spreads depends on the rate of population growth and percentage of the population that disperses. In this exercise, we examine the effects of the growth rate and the dispersal rate on the spread of a pest population, where we attempt to slow the spread, and in an endangered species, where we attempt to speed up the spread.

Spread of a Pest Population

Some populations of insect herbivores occasionally undergo rapid population growth and then spread throughout a habitat. The Green Budworm, for example, occasionally undergoes exponential growth in one region of a spruce-fir forest and then spreads outward, defoliating the conifer trees as the population invades new regions where the trees are still healthy.

The Green Budworm is the larva of a species of moth. The life cycle is completed within a single year: eggs layed in August hatch in May; larvae develop during the summer and metamorphose into adult moths in August. The adults die after they reproduce. Various sources of mortality operate at different stages in the life cycle. The

Population Ecology: An Introduction to Computer Simulations. By Ruth Bernstein.

larvae and pupae are eaten by birds and parasitized by small wasps (parasitoids). By far the largest source of mortality, however, is from bad weather—particularly in May when the tiny larvae cannot withstand cool, wet weather.

In this exercise we simulate the spread of the Green Budworm under conditions of global warming, in which the weather in May becomes warmer and drier. We expect larval survival to improve dramatically, the population growth rate to increase, and the insects to spread rapidly throughout the spruce-fir forests.

Suppose you are a forest biologist and want to slow the rate at which the budworms are spreading. You conclude that the rate of spread depends mainly on the rate of population growth and the proportion of each population that disperses (moves away from where it developed) each year. You need to determine the relative importance of these two parameters in order to develop a successful management program. A computer simulation is the best way to explore such variables. As these insects reproduce once a year, you decide to use the discrete model of exponential growth. Keep in mind that the rate of growth, R, must be greater than 1 for the population to spread.

First, write a program that tells the computer what to do, and then store it so you can use it repeatedly without retyping. Open Microsoft Word and then open MATLAB. Scroll down the "file" icon and then click "new" and "m-file." Notice that the window that opens (called the program window) lacks the >> symbol. Now write a program called *invasion* by typing the following:

```
function edge=invasion(hablen,runlen,estab,no,d,r)
hab=zeros(1,hablen); new_hab=hab;
hab(1)=no; edge=[1];
for t=1:runlen
  for h=1:hablen
    hab(h)=r*hab(h);
  end
  new_hab(1)=((1-d)+d/2)*hab(1)+(d/2)*hab(2);
  for h=2:(hablen-1)
    new_hab(h)=(d/2)*hab(h-1)+(1-d)*hab(h) + (d/2)*hab(h+1);
  end
  new_hab(hablen)=((1-d)+d/2)*hab(hablen)+(d/2)*hab(hablen-1);
  hab=new_hab;
  edge=[edge min(find(hab<estab))];
end
```

Check your program carefully to be sure it is typed correctly. Even a small error will prevent the program from running. Fortunately, it is easy to correct errors in the program window. Store the program with its file name (*invasion*) on a floppy disk by scrolling down "file" to "save as." This set of commands is called into play (but not displayed) whenever you type *invasion* in the command window. Close the program window and return to the command window.

Choose values for the following four parameters: `hablen` is the length of the habitat through which the organisms spread, `runlen` is the number of years you want to simulate, `d` is the proportion of the population that disperses each year, `estab` is the number of individuals needed to establish a new population, and `no` is the number of individuals in the population at the beginning of the simulation. In the example below, I used a habitat length of 100 kilometers (assuming that each adult moth can disperse over about 1 kilometer), a run of 50 years, a requirement of 10 individuals to establish a population, and an initial population size of 100.

In the command window, tell the computer you will be using your floppy disk:

```
>> cd a:
```

Then give the specifics of your run. For my run, I used:

```
>>hablen=100; runlen=50; estab=10; no=50;
>>figure
>>hold on
```

Assign the values of d (the proportion that disperses) and R (the population growth rate for discrete growth), as well as the color or line style that you want to appear on the graph. In my first run, I assumed that 20 percent of the adult moths disperse each autumn ($d = 0.2$) and that global warming will quadruple the normal growth rate (from $R = 1$ for a stable population to $R = 4$). These two values are inserted after the `no` entry. I chose the color green.

```
>>plot(invasion(hablen,runlen,estab,no,0.2,4),'g')
```

Now, on the same graph, try another value of d and plot it in another color or line style. To do this, you need only retype the `plot(invasion(hablen,runlen,estab, no, d, R), 'color or line style')` commands. In this graph, compare only the effects of d on the rate of spread, keeping R constant. Continue to type this plot command for all the variations you want to do.

Begin a new graph, by typing "`figure, hold on`", to explore the effects of R (within a realistic range) on the rate of spread. For this graph, hold d constant and simulate the spread with several different values of R.

Give each graph a title and label the axes (the X axis is time in years and the Y axis is spread of the population, in kilometers). Identify the values of the parameters for each curve. Transfer the graph to a Word document by scrolling down "edit" and clicking "copy figure" and then, in the Word document, moving the cursor to a position somewhat below the upper margin and clicking "paste." Write a figure legend explaining how the simulation was done and how, as a forest biologist, you would reduce the rate of reproduction R or the proportion of adults that successfully disperse (d) in the field. Move the cursor below the figure to position the document for the next graph.

Spread of an Endangered Population

Now suppose that you are a conservation biologist who is planning the reintroduction of an endangered species. What you want to find out, through computer simulations, is the best way to increase the rate at which this population spreads throughout its habitat. Consider here a small group of lynx that you plan to introduce into a spruce-fir habitat. Use the *invasion* program to see how to maximize the success of the lynx restoration program.

Consider two factors that affect the spread of lynx: dispersal rate (d), which is at least 0.5 (i.e., at least half of the lynx disperse each year), and the population growth rate (R), which you know to be at most 1.3. You estimate that at least three lynx are needed to establish a new subpopulation in order to have a high probability of being one female and one male. (For a random sample of three individuals, the probability of having one male and one female is one minus the probability of having all females or all males, which is

$$1-\left[\left(\frac{1}{2}\right)\left(\frac{1}{2}\right)\left(\frac{1}{2}\right)+\left(\frac{1}{2}\right)\left(\frac{1}{2}\right)\left(\frac{1}{2}\right)\right]=1-\frac{2}{8}=0.75$$

Change the growth rate and the dispersal rate systematically, one at a time, to find out which has the greatest effect on the spread of the lynx. Give a title to the graph, label the axes, and identify the values of the parameters used for each curve.

Transfer the graphs to a Word document. Write a figure legend for each graph, describing exactly what you would do in the field to increase the rate of spread.

EXERCISE 3

The Leslie Matrix: Age Structured

A population growing at a constant rate reaches a stable age distribution in which the proportion of individuals in each age class remains the same from one year to the next. The Leslie matrix is an algebraic matrix that is used to predict this stable age distribution and to calculate the population growth rate after this distribution is established. This technique was developed in the late 1930s by the British mathematician Patrick Leslie. A Leslie matrix is constructed from information in a life table, which is a summary of age-specific rates of survival and reproduction. A life table usually consists only of information on the females of the population, because it is hard to keep track of how many offspring a male produces.

Construction of a Life Table from Field Data

Suppose that you are interested in the population dynamics of Silver-back Ground Squirrels. You mark 100 newborn females within a 10 hectare study site, defining a newborn as a young squirrel when it makes its first appearance above ground in the spring. The age of a newborn is considered to be zero. This marked group of individuals, born during the same reproductive season, is known as a cohort. Every spring you return to the site and count the number of marked squirels, which you assume to be the survivors of the newborns marked during the first year of your study. You also count how many females have litters and the number of young within each litter. Your field data are summarized in Table 3.1.

Population Ecology: An Introduction to Computer Simulations. By Ruth Bernstein.

Table 3.1 Field data for the Silver-back Ground Squirrel

Year of study	Age of cohort	Number still alive	Number of female newborns
first	0	100	0
second	1 year old	40	0
third	2 years old	30	60
fourth	3 years old	20	60
fifth	4 years old	20	60
sixth	5 years old	0	0

The field data are used to construct a life table, which consists of the age (x) of the cohort, the survivorship probabilities (l_x), and the fecundity values (m_x). The survivorship probabilities (l_x) are the probabilities of surviving from age 0 to age x. They are calculated by

$$l_x = \frac{\text{number alive at age } x}{\text{number at age } 0} \tag{3.1}$$

For example,

$$l_3 = \frac{20}{100} = 0.2$$

The fecundity function (m_x) is simply the average number of female offspring left by a female of age x. It is calculated by

$$m_x = \frac{\text{total number of female newborns produced by age class } x}{\text{total number of females of age class } x} \tag{3.2}$$

For example,

$$m_3 = \frac{60}{20} = 3$$

The life table for your population of Silver-back Ground Squirrels is shown in Table 3.2.

Table 3.2 Life table for the Silver-back Ground Squirrel

Age x	l_x	m_x
0	1.0	0
1	0.4	0
2	0.3	2
3	0.2	3
4	0.2	3
5	0	0

Calculation of the Population Growth Rates: R_0, r, and R

All three estimates of population growth can be calculated from Table 3.2. These are R_0, the rate of growth per generation; r, the rate of growth per individual in continuous time; and R, the rate of growth per individual in discrete time.

The rate of growth per generation is calculated first. This rate, which is defined as the average number of female offspring left by a female in her lifetime, is known as the net replacement rate and symbolized by R_0:

$$R_0 = \sum l_x m_x \tag{3.3}$$

Use MATLAB to solve for R_0:

```
>> R0=(1*0)+(0.4*0)+(0.3*2)+(0.2*3)+(0.2*3)
```

To calculate the rate of growth for continuous time r, we need to estimate the generation time T, which is the average time between when a female has her offspring and when her daughters have their offspring:

$$T = \frac{\sum l_x m_x x}{\sum l_x m_x} \tag{3.4}$$

Note that the denominator is the same as R_0. Using the value of R_0 calculated above, find T by typing

```
>>T=((0*1.0*0)+(1*0.4*0)+(2*0.3*2)+(3*0.2*3)+(4*0.2*3))/R0
```

To calculate the rate of growth per individual for continuous time (r) from R_0 and T, begin with the equation for exponential growth:

$$N_t = N_0 e^{rt} \tag{3.5}$$

Modify this equation for a specific time period—one generation (T):

$$N_T = N_0 e^{rT}$$

Divide both sides of the equation by N_0:

$$\frac{N_T}{N_0} = e^{rT}$$

N_T/N_0 is the number of individuals in the population one generation from now divided by the number in the population now. This ratio is the same as the rate of growth per generation (R_0). Thus

$$R_0 = e^{rT} \tag{3.6}$$

Taking the natural log of each side and dividing both sides of the equation by T, we get

$$\frac{\ln R_0}{T} = r$$

Find the value of r for the ground squirrels described in the life table. Use the above equation, substituting the values of R_0 and T calculated above:

```
>>r=log(R0)/T
```

Lastly convert the population growth rate r (for continuous growth) into the population growth rate (R) for discrete growth. Begin with the two growth equations, the first for continuous time and the second for discrete time:

$$N_t = N_0 e^{rt} \quad \text{and so} \quad \frac{N_t}{N_0} = e^{rt}$$

$$N_{t+1} = RN_t \quad \text{and so} \quad \frac{N_{t+1}}{N_t} = R$$

If the time period is assigned a value of 1 (which it always is for discrete growth), then

$$\frac{N_t}{N_0} = \frac{N_{t+1}}{N_t}$$

and, since $t = 1$,

$$e^r = R \tag{3.7}$$

Use MATLAB to solve for R using the value of r calculated above:

```
>>R = exp(r)
```

While the population growth rate can be calculated in this manner, directly from a life table, it is less accurate than the rate calculated from the Leslie matrix.

Construction of a Leslie Matrix from the Life Table

An age-structured matrix describes rates of survivorship and fecundity for each age of the organisms in a population. The units of age may be in days, weeks, months, or years, depending on the lifespan of the organism. For the matrix, we need to calculate the probability of surviving from one age to the next (p_x) and the fecundity of females per year (F_x) for each age class. Each p_x value is the probability that an individual will survive from age x to age $x + 1$, calculated by

$$p_x = \frac{l_{x+1}}{l_x} \tag{3.8}$$

For example,

$$p_2 = \frac{l_3}{l_2}$$

Each F_x value is a combination of m_x and the probability of living from last year (age $x - 1$) to the present year (age x), which is p_{x-1}. We assume here that the population is censused after the reproductive season, so the newborns next year are from females who survived from this year to next year. The fecundity of a female of age x, then, is the probability that she survived from $x - 1$ to age x multiplied by the average fecundity of x-year-old females (m_x):

$$F_x = p_{x-1} m_x \tag{3.9}$$

For example,

$$F_3 = p_2 m_3$$

To estimate the number of individuals that will be in each age category next year, we multiply the Leslie matrix by the number of individuals in each age category this year. Thus, the number of individuals in each age category (except newborns) next year is the number in that age category this year times their probability of surviving to the next year:

$$n_{x+1} = n_x p_x$$

The p_x values are incorporated within the Leslie matrix. For example, the number of 1-year-olds next year will be the number of newborns this year times their probability of surviving from age 0 to age 1.

To estimate the number of newborns in the population next year, we calculate the number of offspring produced by all females in the population next year. Fortunately, the Leslie matrix does this for us, given the values of p_x and F_x for each age x and the initial number of individuals in each age category.

We now convert this information into a Leslie matrix, which is always a square matrix with the number of rows equal to the number of columns equal to the number of age categories (including the newborns). Our ground squirrel population has five age categories: 0, 1, 2, 3, and 4 (all 5-year-olds are dead), giving us a 5 by 5 matrix. The generic version of this matrix is shown in Table 3.3. Construct a Leslie matrix, using real values of p_x and F_x for the ground squirrel population. *Throughout these instructions, I apply numbers that are not from this particular Leslie matrix.* You should substitute values from your own Leslie matrix for ground squirrels.

Table 3.3 Leslie matrix for the Silver-backed Ground Squirrel: generic version

F_1	F_2	F_3	F_4	0
p_0	0	0	0	0
0	p_1	0	0	0
0	0	p_2	0	0
0	0	0	p_3	0

Application of the Leslie Matrix

Transplant a small group of ground squirrels from your population to an "empty" habitat (in which the previous colony was wiped out by disease) and see what happens to the age distribution of this transplanted group through time. In my study, I transfer 10 1-year-old females, and 50 2-year-old females (and enough males to get all the females mated) from a crowded area to this "empty" habitat. I will count only the females.

We are now ready to use MATLAB. First, describe your matrix (five rows with five entries each) to the program. This can be done on a single line, with semicolons between the rows, or it can be done by typing "`enter`" after each semicolon. (Remember, the numbers below are not the ones you should use.)

```
>>m = [0 1.8 2 2 0;
      .6 0 0 0 0;
      0 .5 0 0 0;
      0 0 .4 0 0;
      0 0 0 .2 0]
```

Then, describe the transplanted group—the numbers of animals in each of the five age classes (`n0`). (Write down these numbers on a piece of paper so you have a record of what you did.)

```
>>n0 = [0; 10; 50; 0; 0]
```

Next, find the proportion of individuals in each age class. Tell the computer to do this by summing the number of individuals in all age classes (s) and then dividing the number in each age class (n) by this sum (i.e., n/s). Begin with time 0:

```
>>s0 = sum(n0);
>>c0 = n0/s0;
```

Then, determine the number of individuals in the population next year (`n1`), based on the p_x and F_x values in the matrix (`m`):

```
>>n1=m*n0;
>>s1=sum(n1);
>>c1=n1/s1;
>>n2=m*n1;
>>s2=sum(n2);
>>c2=n2/s2;
```

Continue this process through `c8`, omitting the semicolon after `c8` in order to have the program tell you the age distribution at that point in time.

You have now generated eight iterations (eight years of population growth) and have a total of nine population sizes and nine age distributions, including the initial group that you transferred into the "empty" habitat. To see what has happened to your population of ground squirrels in its new habitat, have the computer draw graphs of the nine age distributions. They can all be arranged within one figure, with nine separate panels. The `subplot(m,n,r)` command makes the next plot occur in the rth panel of an m by n array of graphs (here, a 3 by 3 array, for the nine years). To get this graph, type

```
>>figure
>>subplot(3,3,1)
>>bar(c0)
>>subplot(3,3,2)
>>bar(c1)
```

Continue this procedure until you reach

```
>>subplot(3,3,9)
>>bar(c8)
```

The sequence of graphs is from left to right along the first row, then from left to right along the second row, etc. Identify each graph in your figure by clicking on the first graph (to form a frame) and clicking "A" on the toolbar. Then click a spot within the figure and type "`year 0`." Click on an empty place in the figure to fix the label. Next, click on the second figure (to form a frame), etc. The results are more easily interpreted when all the *Y* axes are the same; if they are not all the same, adjust each one to your highest value by clicking on the figure (to form a frame), scrolling down "edit" on the toolbar, clicking "axes properties." In this window, click "auto" just before the *Y*, adjust the axis, and then click "OK" at the bottom of the window. As it is difficult to label the *X* and *Y* axes of all the graphs, simply identify the axes in your figure legend. (The *Y* axes are the proportions of the population in each age category and the *X* axes are the age categories, indicated by the maximum age within each category, e.g., age 0 to 1 is indicated as age 1.) Transfer your figure to a Microsoft Word document. Write a figure legend describing what happened to the age distribution of your population through the years.

To find out how close your population is to a stable age distribution, have MATLAB calculate the age distributions after many years (e.g., 20) and compare this distribution with the one you found above (after the `c8` command).

```
>>n20=m^19;
>>s20=sum(n20);
>>c20=n20/s20
```

To graph this age distribution, type

```
>>bar(c20)
```

See what happens to the size of your transplanted population by graphing the sequence of total population sizes through time. Enter the time interval (from time 0 to time 8, in one year intervals) and the total size vectors (*s* values):

```
>>t=0:1:8;
>>s=[s0 s1 s2 s3 s4 s5 s6 s7 s8];
```

Construct the graph with these commands:

```
>>figure
>>plot(t,s)
```

Label this graph and transfer it to your Word document. Give it a figure legend. How does this graph help you to interpret the rates of growth observed the first few years after a population is introduced into a new environment?

Now have the program find the growth rate, R, for your transplanted ground squirrels. Recall that the discrete growth equation is

$$N_{t+1} = RN_t \tag{3.10}$$

The Leslie matrix calculates N_{t+1} for each age class. Once a stable age distribution is reached, the proportion of individuals within each age class remains the same, but the number of individuals changes at the same rate, R. The number of individuals in each age class at time $t + 1$ is the number at time t multiplied by the algebraic solution to the matrix:

$$N_{t+1} = mN_t \tag{3.11}$$

where m is the Leslie matrix. As you can see from Equations (3.10) and (3.11), m is the same as R. The matrix, however, is a complex algebraic formula and so has more than one solution. Each solution is called an eigenvalue. The most important of these eigenvalues is the largest one, because it eventually dominates the others. This largest, dominant eigenvalue is the discrete rate of population growth R. To find this value, have the program find the absolute values of all eigenvalues and then find the largest of these absolute values. To do this, enter

```
>>R=max(max(abs(eig(m))))
```

Compare the R value that you obtained from the Leslie matrix with the one that you calculated directly from the life table. The R value calculated from the life table is less accurate because the generation time T is an average, whereas it is not an average in the Leslie matrix.

EXERCISE 4

The Leslie Matrix: Stage Structured

A population growing at a constant rate reaches a stable age distribution in which the proportion of individuals in each age class remains the same from one year to the next. The Leslie matrix is used to predict this stable age distribution and to calculate the population growth rate after this distribution is established. This technique was developed by Patrick Leslie, a British mathematician, around 1940.

A Leslie matrix is constructed from information in a life table. It is not always possible, however, to obtain a complete table with survivorship and fecundity functions for each age class. A stage-structured matrix has at least some of the functions for a stage (rather than an age) of life. It is used when individuals are censused (rather than marked from birth) and the age of each individual cannot be known for certain. In such populations, individuals are described by size class (e.g., fishes), physical attributes (e.g., deer), or by stage of life (e.g., frogs) rather than by specific ages.

In this exercise, you will construct two stage-structured matrices, one for a population of moths and the other for a population of elk. In order to evaluate the accuracy of the stage-structured matrix, we will pretend that a complete age-structured life table exists for each population and calculate from this table an estimate of R.

Population Ecology: An Introduction to Computer Simulations. By Ruth Bernstein.

Calculation of the Population Growth Rates from an Age-structured Life Table

All three estimates of the population growth rate can be calculated from a life table. These are R_0, the rate of growth per generation; r, the rate of growth per individual in continuous time; and R, the rate of growth per individual in discrete time.

The rate of growth per generation is calculated first. This rate, which is defined as the average number of female offspring left by a female in her lifetime, is known as the net replacement rate and symbolized by R_0:

$$R_0 = \sum l_x m_x \tag{4.1}$$

To calculate the rate of growth for continuous time (r), we need to estimate the generation time T, which is the average time between when a female has her offspring and when her daughters have their offspring:

$$T = \frac{\sum l_x m_x x}{\sum l_x m_x} \tag{4.2}$$

To calculate the rate of growth per individual for continuous time (r) from R_0 and T, begin with the equation for exponential growth:

$$N_t = N_0 e^{rt} \tag{4.3}$$

Modify this equation for a specific time period – one generation (T)

$$N_T = N_0 e^{rT} \tag{4.4}$$

Divide both sides of the equation by N_0:

$$\frac{N_T}{N_0} = e^{rT}$$

N_T/N_0 is the number of individuals in the population one generation from now divided by the number in the population now. This ratio is the same as the rate of growth per generation (R_0). Thus

$$R_0 = e^{rT} \tag{4.5}$$

Taking the natural log of each side and dividing both sides of the equation by T, we get

$$\frac{\ln R_0}{T} = r$$

Lastly the population growth rate r (for continuous growth) can be converted into the population growth rate (R) for discrete growth. Begin with the two growth equations, the first for continuous time and the second for discrete time:

$$N_t = N_0 e^{rt} \quad \text{and so} \quad \frac{N_t}{N_0} = e^{rt} \tag{4.6}$$

$$N_{t+1} = RN_t \quad \text{and so} \quad \frac{N_{t+1}}{N_t} = R \tag{4.7}$$

If the time period is assigned a value of 1 (which it always is for discrete growth), then

$$\frac{N_t}{N_0} = \frac{N_{t+1}}{N_t}$$

and, since $t = 1$,

$$e^r = R \tag{4.8}$$

Green-spotted Moth

The Green-spotted Moth is a tropical moth that passes from the egg stage through the caterpillar and pupa stages to the adult stage in four weeks. A stage-structured Leslie matrix is used for this moth because the caterpillar stages are difficult to age. While an age-structured matrix would be more accurate, the detailed information needed for describing each age in a population is not available. Here, we pretend that an age-structured table is available, in order to evaluate the stage-structured results. Assume, then, that the moth population has the age-structured life table shown in Table 4.1.

Table 4.1 Age-structured life table for the Green-spotted Moth

Age x (in weeks)	l_x	p_x	m_x
0 (egg)	1.000	0.5	0
1 (caterpillar)	0.500	0.5	0
2 (caterpillar)	0.250	0.5	0
3 (pupa)	0.125	0.8	0
4 (adult)	0.100	0	125

The average number of female offspring left by a female in her lifetime (in weeks), symbolized by R_0, is calculated by using Equation (4.1). Use MATLAB to calculate R_0 from the data in Table 4.1:

```
>>R0=(1*0)+(0.5*0)+(0.25*0)+(0.125*0)+(0.1*125)
```

As each moth reproduces just once, at four weeks of age, the generation time T is four weeks. Calculate r and then R (per week):

```
>>r=log(125)/4
>>R=exp(r)
```

Now, assume that you cannot determine the age, in weeks, of the moth but you can distinguish among the egg, caterpillar, pupa, and adult stages. Thus, you use these stages to construct a stage-structured matrix. First, convert the age-structured life table into a stage-structured life table, as shown in Table 4.2. Then construct a stage-structured Leslie matrix that includes the following information:

$F_x = p_{x-1}m_x$ Here, there is only one reproductive stage – the adult moth – and F_{adult} is equal to (0.8)(125), or 100.

$p_{\text{caterpillar}}$ = probability that a caterpillar becomes a pupa, which is calculated by multiplying the (probability that a caterpillar survives from one week to the next) by (the proportion of caterpillars that become pupae).

G = probability that a caterpillar remains a caterpillar, which is calculated by multiplying the (probability of surviving from one week to the next while in the caterpillar stage) by (the proportion of caterpillars that remain as caterpillars rather than moving on to the pupa stage).

The generic version of this Leslie matrix for the Green-spotted Moth is shown in Table 4.3.

To find these values for the Leslie matrix, assume that reproduction is continuous and that each series begins with 100 eggs. Applying the survivorship values, the numbers of

Table 4.2 Stage-structured life table for the Green-spotted Moth

x	Time spent in stage	p_x*	m_x
egg	1 week	0.5	0
caterpillar	2 weeks	0.5	0
pupa	1 week	0.8	0
adult	1 week	0	125

*probability of surviving from one week to the next

Table 4.3 Stage-structured Leslie matrix for the Green-spotted Moth: generic version

0	0	F_{adult}	0
p_{eggs}	G	0	0
0	$p_{caterpillars}$	0	0
0	0	p_{pupae}	0

individuals in each category are then: 100 eggs, 50 + 25 caterpillars, 12.5 pupae, and 10 adults. Each week, 25 of the 75 caterpillars (0.333) are ready to become pupae. The probability of surviving this transition is 0.5. Thus, the probability that a caterpillar becomes a pupa ($p_{caterpillar}$) is (0.333)(0.5), or 0.167. Each week, 50 of the 75 caterpillars (0.677) remain as caterpillars and their probability of living to the next week is 0.5. Thus, the probability that a caterpillar will remain as a caterpillar for another week (G) is (0.677)(0.5), or 0.333. The stage-structured matrix for this moth is shown in Table 4.4.

Table 4.4 Stage-structured Leslie matrix for the Green-spotted Moth

0	0	100	0
0.5	0.333	0	0
0	0.167	0	0
0	0	0.8	0

Enter this matrix in MATLAB and then have the program calculate R:

```
>>m = [ 0 0 100 0;
      .5 .333 0 0;
      0 .167 0 0;
      0 0 .8 0 ]
>>R=max(max(abs(eig(m))))
```

How does this value of R compare with the value that you calculated directly from the life table? What are the sources of error?

Have the program calculate the stable stage distribution after 20 generations. Be sure to omit the semicolon after the `c20` command, as these values are the proportions of individuals in each stage category.

```
>>n20 = m^19;
>>s20 = sum(n20);
>>c20 = n20/c20
```

Graph this stable stage distribution.

```
>>bar(c20)
```

Tawny Elk

Suppose that you find, in the biological literature, the life table shown in Table 4.5 for a population of Tawny Elk. In this table, x is the age in years (with age 0 equal to a newborn), l_x is the probability of living from age 0 to age x, p_x is the probability of living from age x to age $x + 1$ (calculated by dividing l_{x+1} by l_x), and m_x is the average number of female offspring left by a female of age x. As you can see, a cow begins to reproduce when she is 4 years old. She then has one offspring per year (i.e., 0.5 female offspring) every year until she dies. To find the rate of population growth, you need first to find R_0, which is the average number of female offspring left by a female in her lifetime. Applying Equation (4.1), you find that R_0 is equal to 1.25. Applying Equation (4.2), you find that the

Table 4.5 Life table for the Tawny Elk

x	l_x	p_x	m_x
0	1.00	0.71	0
1	0.71	0.75	0
2	0.53	0.77	0
3	0.42	0.79	0
4	0.33	0.94	0.5
5	0.31	0.90	0.5
6	0.28	0.93	0.5
7	0.26	0.92	0.5
8	0.24	0.92	0.5
9	0.22	0.95	0.5
10	0.21	0.86	0.5
11	0.18	0.89	0.5
12	0.16	0.94	0.5
13	0.15	0.40	0.5
14	0.06	0.42	0.5
15	0.025	0.42	0.5
16	0.011	0.45	0.5
17	0.005	0	0.5

generation time, T, is equal to 8.236. Use MATLAB to calculate the continuous rate of population growth:

$$r = \frac{\ln R_0}{T}$$

that is,

```
>>r=log(1.25)/8.236
```

and the discrete rate of population growth:

$$R = e^r$$

that is,

```
>>R=exp(r)
```

This R is only an estimate of the real rate of growth per year because generation time is an average. To find the real rate, you need to find the R value obtained from an age-structured Leslie matrix. As an age-structured life table for your population of Tawny Elk is not, in fact, available (we imagined here that it was, for purposes of comparison), you use instead a stage-structured table.

Construction of the stage-structured matrix

In your study of the Tawny Elk, you are able to distinguish among the newborns, 1-, 2-, and 3-year-olds, but for the adult cows you can distinguish only those that are elderly (ages 13 through 17) from those that are middle aged. Thus, in this research project, you group the adult females into two stages – middle-aged and elderly. These groupings are not unrealistic, since the p_x values and the m_x values for the middle-aged cows are very similar (average $p_x = 0.92$; all $m_x = 0.5$) as are the p_x values and the m_x values for the elderly cows (average $p_x = 0.42$; all $m_x = 0.5$).

To construct a stage-structured Leslie matrix, it is useful to construct a flow diagram of the movement of individuals through a life table. Such a diagram is shown for the Tawny Elk population in Figure 4.1. Each box in the diagram represents either an age category or a stage category. The symbols used in this diagram are:

G_m : the probability that a middle-aged adult lives to the next year (0.92) multiplied by the proportion of middle-aged adults that remain in that stage. Before the "birthdays" of the elk, stage m consists of individuals from age 4 through age

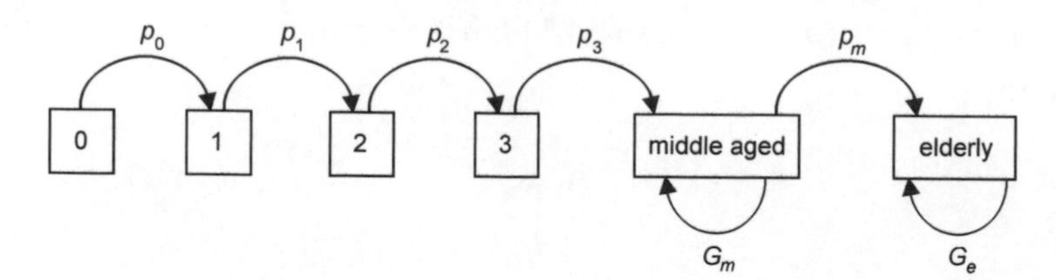

Figure 4.1 Flow diagram for organizing a stage-structured Leslie matrix for the Tawny Elk. Symbols: p_x = probability of surviving from one age to the next; G_m = probability of remaining in the middle-aged adult stage = (probability of surviving from one age to the next as a middle-aged adult) × (probability of remaining in the middle-aged stage rather than move to the elderly stage); G_e = probability of remaining in the elderly adult stage = (probability of surviving from one age to the next as an elderly adult) × (probability of remaining in the elderly stage rather than die)

12; after their "birthdays," stage m consists of the same individuals except the 12-year-olds, who have moved into the elderly group. Thus, the proportion of individuals that remain in the group is the number of elk of ages 4 through 11 divided by the number of elk of ages 4 through 12. The proportion can be obtained by substituting the l_x values for the numbers of elk. Thus,

$$\text{proportion that remains in stage } m = \frac{\sum_{x=4}^{11} l_x}{\sum_{x=4}^{12} l_x} \tag{4.9}$$

$$\text{proportion that remains in stage } m =$$

$$\frac{0.33 + 0.31 + 0.28 + 0.26 + 0.24 + 0.22 + 0.21 + 0.18}{0.33 + 0.31 + 0.28 + 0.26 + 0.24 + 0.22 + 0.21 + 0.18 + 0.16} = 0.927$$

And so, $G_m = (0.92)(0.927) = 0.853$.

p_m : the probability that a middle-aged adult will survive to the next year and enter the elderly stage. It is equal to the probability of surviving from one year to the next (0.92) times the proportion that leave the middle-aged stage, which is 1 minus the proportion that stay in the middle-aged stage:

$$p_m = 0.92(1 - \text{proportion that stay}) = 0.92(0.073) = 0.067$$

G_e : the probability of an elderly elk living from one year to the next (p_e) multiplied by the proportion that remain in stage e (rather than die). Thus

$$\text{proportion that remains in stage } e = \frac{\sum_{x=13}^{16} l_x}{\sum_{x=13}^{17} l_x} = \frac{0.14 + 0.09 + 0.05 + 0.02}{0.14 + 0.09 + 0.05 + 0.02 + 0.01} = 0.968$$

$$G_e = (0.42)(0.968) = 0.407$$

The generic version of your stage-structured Leslie matrix is shown in Table 4.6.

Table 4.6 Stage-structured Leslie matrix for the Tawny Elk: generic version

F_1	F_2	F_3	F_4	F_m	F_e
p_0	0	0	0	0	0
0	p_1	0	0	0	0
0	0	p_2	0	0	0
0	0	0	p_3	G_m	0
0	0	0	0	p_m	G_e

When this matrix is multiplied by the number of newborns, 1-, 2-, and 3-year olds, middle-aged adults, and elderly adults in the population this year, the result is the number of individuals in each age or stage category next year.

The number of newborns is the sum of the fecundity functions of each stage:

$$F_1 = p_0 m_1 = (0.71)(0) = 0$$
$$F_2 = p_1 m_2 = (0.75)(0) = 0$$
$$F_3 = p_2 m_3 = (0.77)(0) = 0$$
$$F_4 = (p_3 m_4) = (0.79)(0.5) = 0.395$$
$$F_m = (p_m m_5) + (p_m m_6) + (p_m m_7) + (p_m m_8) + (p_m m_9) + (p_m m_{10}) + (p_m m_{11}) + (p_m m_{12})$$
$$= (0.92)(0.5)(8) = 3.68$$
$$F_e = (p_m m_{13}) + (p_e m_{14}) + (p_e m_{15}) + (p_e m_{16}) + (p_e m_{17})$$
$$= (0.92)(0.5) + (0.42)(0.5)(4) = 1.30$$

The stage-structured matrix for your Tawny Elk population is shown in Table 4.7.

Table 4.7 Stage-structured Leslie matrix for the Tawny Elk

0	0	0	0.395	3.68	1.30
0.71	0	0	0	0	0
0	0.75	0	0	0	0
0	0	0.77	0	0	0
0	0	0	0.79	0.853	0
0	0	0	0	0.067	0.407

Application of the matrix

Have MATLAB calculate R by entering the matrix:

```
>> m = [0 0 0 .395 3.68 1.30;
        .71 0 0 0 0 0;
        0 .75 0 0 0 0;
        0 0 .77 0 0 0;
        0 0 0 .79 .853 0;
        0 0 0 0 .067 .407]
>>R = max(max(abs(eig(m))))
```

Ask MATLAB to calculate the stage distribution after 20 years. (Be sure to omit the semicolon after the last command, so the program will give you the proportions of individuals in each stage category.)

```
>>n20 = m^19;
>>s20 = sum(n20);
>>c20 = n20/s20
```

This list of proportions is the stable stage distribution for your elk population. Graph it by typing

```
>>bar(c20)
```

Label the graph and transfer it to a Microsoft Word document.

Compare the two estimates of R, one from the life table and one from the matrix. Which measure do you think is more accurate? Why? The most accurate measure, of course, would be an age-structured matrix in which a large number of newborns are marked at birth and followed until death. An age-structured matrix for this Tawny Elk population generates an R of 1.01.

EXERCISE 5

Metapopulation Dynamics

A metapopulation is a population in which the individuals are spatially distributed within the habitat as two or more subpopulations interconnected by dispersal. Natural populations of butterflies and coral reef fishes, for example, typically form metapopulations because individuals occupy patches of suitable habitat. Human actitivies are increasing the number of populations that occur as metapopulations; such activities fragment large areas of continuous habitat into smaller patches of habitat. For this reason, models of metapopulation dynamics have become important tools in the field of conservation biology.

The Levins Model

The concept of a metapopulation was introduced in 1969 by Richard Levins, an American population ecologist. The Levins model is based on a population in which individuals reproduce and die within local patches of the habitat, and their offspring disperse into other patches. The number of individuals within each patch fluctuates greatly, so that the subpopulation within a patch is vulnerable to extinction. The model is written in the form of a differential equation:

Population Ecology: An Introduction to Computer Simulations. By Ruth Bernstein.

$$\frac{dp}{dt} = c(1 - p)p - ep \tag{5.1}$$

where c is the rate at which an occupied patch produces colonists, p is the proportion of patches that are occupied, and $1 - p$ is the proportion of patches that are vacant. Thus, $cp(1 - p)$ is the rate at which vacant patches become occupied patches. The rate at which occupied patches become vacant patches is the probability that a subpopulation within a patch goes extinct (e) times the proportion of patches that are occupied (p). The model assumes that (1) the metapopulation exists within a homogeneous habitat that is subdivided into patches and (2) the young disperse randomly to all possible patches within the habitat. This model, while simple, forms the foundation of all later work on metapopulation dynamics.

Consider a population of Smoky Butterflies that lives on Goldenbush, a host plant that occurs in moist patches within a scrub habitat. To make the model specific to this population, define p as the *number* of Goldenbush patches occupied by the butterfly and h as the *total number* of Goldenbush patches present in the habitat. Thus, $h - p$ is the number of vacant Goldenbush patches. As in the equation above, c is the rate at which an occupied patch produces colonists and e is the rate at which an occupied patch goes extinct:

$$\frac{dp}{dt} = c(h - p)p - ep \tag{5.2}$$

Ask MATLAB to solve the differential equation:

```
>> p=simplify(dsolve('Dp=c*(h-p)*p-e','p(0)=po','t'))
```

Then assign values to `h`, `e`, `c`, and `po` (the initial number of occupied patches) and set a time period. I began my exploration of the model by setting the number of Goldenbush patches at 10, an extinction probability of 0.4 each year, and a colonization probability of 0.1 each year. For the first simulation, I assumed that some catastrophic event had wiped out all subpopulations but one, therefore `po` was set at 1.

```
>> h=10; e=0.4; c=0.1;
>> po=1;
>> t=0:1:20;
>> figure
>> hold on
>> plot(t,eval(vectorize(p)),'r')
```

The X axis is years and the Y axis is number of Goldenbush patches occupied by the Smoky Butterfly. Adjust the Y axis to go from 0 to 10 (by scrolling down "edit" to "axes properties"). Now, try different values of `po`, `e`, and `c`, changing just one variable at a time. To do this, you only have to define the new value (e.g., `po=10`) and then type the `plot` command. See if you can establish an equilibrium number of occupied patches in which all patches are occupied. How does this model explain real observations of nature in which suitable patches of habitat contain no individuals of a population? Label the graph, move it to a Microsoft Word document, and write a figure legend.

Metapopulation Stability

The advantage of a metapopulation is its long-term stability. When a population is divided into subpopulations within a heterogeneous environment, the population growth rates of the subpopulations will vary in accordance with local conditions. The metapopulation as a whole, however, may be stable because of dispersal from subpopulations that are thriving into subpopulations that are declining.

There are two ways to look at the effects of local environments on metapopulation stability. In the first way, the quality of one local environment (with regard to the population under study) is in no way correlated with the quality of another local environment – the population growth rates of the subpopulations are independent of one another. Knowing the growth rate of one subpopulation tells you nothing about what is happening in another subpopulation. In the second way, the quality of one local environment is negatively correlated with the quality of another local environment. For example, if the weather that year is unusually warm, then environments on north-facing slopes will be better than usual and environments on south-facing slopes will be worse than usual.

Consider a population divided into subpopulations, each reproducing under a unique set of environmental conditions and so each with a different rate of growth. Assume that reproduction is seasonal, with growth occurring in discrete time. For two such subpopulations, interconnected by dispersal, the discrete equations of growth are

$$N_{1,t+1} = R_{1,t}((1-d)N_{1,t} + dN_{2,t}) \quad (5.3)$$

$$N_{2,t+1} = R_{2,t}((1-d)N_{2,t} + dN_{1,t}) \quad (5.4)$$

where d is the probability that an individual will disperse out of its subpopulation and into the other. Therefore, $1 - d$ is the probability that it will not disperse. (When d is equal to or greater than 0.5, the two subpopulations are completely mixed; they are a single population rather than two subpopulations.)

You will need a new program for this simulation. Open the program window by scrolling down the "file" icon to "new, m-file." In this window, write the following program:

```
function[n1,n2]=meta(d,rgood,rbad,no,runlen,independ)
rand('seed');
n1=[no]; n2=[no];
for t=1:runlen
  if(rand<1/2)
    r1=rbad;
  else
    r1=rgood;
  end;
  if independ==1
    if(rand<1/2)
      r2=rbad;
    else
      r2=rgood;
    end;
  else
    if r1==rgood
      r2=rbad;
    else
      r2=rgood;
    end;
  end;
  n1=[n1 (r1*((1-d)*n1(t)+d*n2(t)))];
  n2=[n2 (r2*((1-d)*n2(t)+d*n1(t)))];
end;
```

(Note that in the bracketed equations, there is a space between the `n1` and `(r1` and between the `n2` and `(r2`.) Save the program as *meta* on your floppy disk. Exit the program window. With this program, you can vary dispersal (abbreviated d) between the sub-populations as well as the degree to which the environments of the two subpopulations vary with regard to each other (abbreviated by *independ*).

Simulate population growth of a metapopulation consisting of two subpopulations in different local environments. Run some of the simulations twice (using the exact same variables) to see the effect of drawing randomly from good and bad years. Such a difference in pattern, when caused by the unpredictable effects of natural variation, is known as a stochastic effect. Here, the "natural" variation is in the drawing of good and bad years from a random number generator. Use a run length of 199 and R values that differ from one but together have an average of 1. (The average density may decrease through time in spite of setting the metapopulation R at 1. This is because the average R over time is a geometric average, which is smaller than the arithmetic average. The geometric average is calculated by $\bar{R}_g = \sqrt[n]{R_1 R_2 R_3 R_n} \ldots$) Begin by telling the program to work from the floppy disk:

```
>> cd a:
>> no=50;
>> runlen=199;
>> rgood=1.25;
>> rbad=0.75;
```

First, do a *control run*, in which the subpopulations have independent population growth rates and no dispersal (i.e., they are not metapopulations). When you invoke the *meta* program (below), the first number in the parentheses after *meta* is the dispersal rate (the proportion of the subpopulation that disperses); the last number is either 1 for no correlation between R values or 0 for a negative correlation.

```
>> [n1,n2]=meta (0,rgood,rbad,no,runlen,1);
>> figure
>> hold on
>> plot(n1,'c')
>> plot(n2,'k')
```

On this graph, X is time and Y is density of the subpopulation. Label the graph, transfer the graph to a Word document, and write a legend. How do these two subpopulations differ? Why?

Now, do an *experimental run* to see the effect of dispersal. I have connected the two populations with a dispersal rate (d) in which 25 percent of each subpopulation disperses.

```
>> [n1,n2]=meta(0.25,rgood,rbad,no,runlen,1);
>> figure
>> hold on
>> plot (n1,'c')
>> plot(n2,'k')
```

Label the graph and transfer it to the Word document. Write a figure legend. What is the effect of dispersal on population stability? What does this tell you about designing wilderness areas for endangered species?

Next examine the effect of environments that are negatively correlated with regard to quality. Again, begin with a control in which there is no dispersal (i.e., the subpopulations do not form a metapopulation). To make the growth rates negatively correlated, simply convert the one to a zero in the last term within the parentheses.

```
>> [n1,n2]=meta(0,rgood,rbad,no,runlen,0);
>> figure
>> hold on
>> plot(n1,'c')
>> plot(n2,'k')
```

Label the graph, transfer it to the Word document, and write a legend. What is the effect of environments that are negatively correlated, with regard to quality, on the population stability of the subpopulations of a metapopulation?

Now, simulate the same subpopulations connected by dispersal (the first term in the parentheses after `meta`):

```
>> [n1,n2]=meta(0.25,rgood,rbad,no,runlen,0);
>> figure
>> hold on
>> plot(n1,'c')
>> plot(n2,'k')
```

Label the graph, transfer it to the Word document, and write a legend. What is the effect of dispersal on subpopulations in environments that are negatively correlated? How does a negative correlation of the *R* values affect the dynamics of subpopulations?

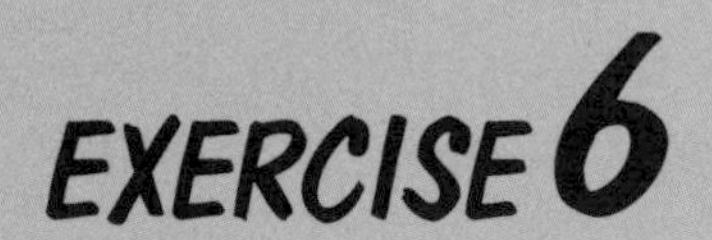

Logistic Population Growth

Populations of organisms do not continue to grow exponentially. Instead, the density increases until it reaches a maximum sustainable density that is set by the availability of resources. This upper limit to population growth, called the carrying capacity of the environment for the population and symbolized as K, was first described in 1845 by the Belgian mathematician Pierre-François Verhulst and then again in 1920 by the American demographers Raymond Pearl and Lowell Reed. Population growth with a carrying capacity is called logistic population growth. (Logistic means to calculate, or to predict from an equation.) It can be modeled using either continuous time or discrete time.

Logistic Population Growth: Continuous Time Model

The continuous time model of population growth is appropriate for populations, such as humans, in which births occur throughout the year rather than during a distinct season. The differential equation for logistic growth is developed from the differential equation for exponential growth by changing the rate of growth per individual from a constant to a variable. In the logistic equation, the rate of growth per individual decreases from a maximum (r_m) to zero as the population grows toward carrying capacity. There are two versions of the model, the classic version and the theta version; they differ in the way in which crowding affects the actual rate of growth per individual.

Population Ecology: An Introduction to Computer Simulations. By Ruth Bernstein.

Classic version

An important assumption of the classic logistic model is that the actual rate of growth per individual (r_a) is a linearly decreasing function of N. To visualize this relationship, enter

```
>>ra='rm*(1-(n/k))';
>>rm=0.6;
>>k=100;
>>n=0:1:110;
>>figure
>>plot(n,eval(vectorize(ra)),'k')
```

The X axis is population density, with the carrying capacity (K) equal to 100, and the Y axis is the actual rate of population growth per individual (r_a). At very low densities, the actual rate of growth is equal to the maximum rate of growth (r_m), which we have assigned a value of 0.6; as density increases, the actual rate of growth declines in a linear fashion. When $N = K$, the rate of growth is zero; when N is greater than K, the rate of growth is negative. Label the graph, move it to a Microsoft Word document, and write a figure legend.

The logistic equation is developed from the exponential equation by substituting a variable $r(r_a)$ for a constant r, with the value of r_a decreasing in a linear fashion with N as shown in your first graph. To develop an equation for this relation between the actual rate of growth and the population density, begin with the equation for a straight line:

$$Y = a + bX \tag{6.1}$$

where a is the Y intercept (r_m here) and b is the slope of the curve. The slope of this line is then calculated by

$$b = \frac{Y_2 - Y_1}{X_2 - X_1} = \frac{0 - r_m}{K - 0} = -\frac{r_m}{K}$$

and so the equation for the relationship between r_a and N is

$$r_a = r_m - \left(\frac{r_m}{K}\right)N \tag{6.2}$$

$$r_a = r_m\left(1 - \frac{N}{K}\right) \tag{6.3}$$

We now substitute this equation, in which r decreases as a function of N, for the constant r in the equation for exponential growth:

$$\frac{dN}{dt} = rN \tag{6.4}$$

$$\frac{dN}{dt} = r_m\left(1 - \frac{N}{K}\right)N$$

which is usually written

$$\frac{dN}{dt} = r_m N\left(1 - \frac{N}{K}\right) \tag{6.5}$$

In MATLAB this equation is entered as `Dn=rm*n*(1-n/k)`. To see how a population grows according to this logistic equation, first ask the program to solve the differential equation

```
>>n=simplify(dsolve('Dn=rm*n*(1-(n/k))','n(0)=no','t'))
```

Then assign values to the variables (I chose an initial density of 10, a maximum r of 0.6, and a carrying capacity of 100) and enter commands for construction of the graph.

```
>>no=10;
>>rm=0.6;
>>k=100;
>>t=0:1:25;
>>figure
>>plot(t,eval(vectorize(n)),'r')
```

The X axis is time and the Y axis is population density. Note that the increase in numbers through time forms an S-shaped curve, with the most rapid increases in N at intermediate densities. Label the graph, move it to a Word document, and write a figure legend.

Theta version

The classic logistic model assumes a straight-line relationship between the actual rate of growth and the population density (see your first graph in this exercise). In other words, each addition to the population causes the same *absolute* decline in growth rate (r_a), regardless of population density. This assumption may be true for some populations, but not for most.

The classic logistic equation is easily modified to adjust for different relationships between the actual rate of growth per individual and the population density. In this modification, known as the theta logistic model, the actual rate of growth is described by

$$r_a = r_m\left[1 - \left(\frac{N}{K}\right)^{\theta}\right] \tag{6.6}$$

which in our computer language is `ra = rm*(1-(n/k)^theta)`. The value of `theta` depends primarily on the relation between crowding and population density, which in turn depends on the pattern of dispersion (the spatial distribution of individuals). In general, where individuals are randomly distributed, `theta` is somewhat larger than 1. Where they are more evenly spaced, `theta` is greater than 2 because crowding has a smaller effect at low densities than at high densities. Where individuals are more aggregated at lower densities than at higher densities, `theta` is less than 1. Graph the relationship between r_a and N for different values of `theta`. (I have used the same values for r_m and K as we used in the classic version, and have chosen values 1, 2, 3, and 0.3 for theta.)

```
>>ra='rm*(1-(n/k)^theta)';
>>rm=0.6;
>>k=100;
>>theta=1;
>>n=0:1:110;
>>figure
>>hold on
>>plot(n,eval(vectorize(ra)),'k')
>>theta=2;
>>plot(n,eval(vectorize(ra)),'g')
>>theta=3;
>>plot(n,eval(vectorize(ra)),'r')
```

```
>>theta=0.3;
>>plot(n,eval(vectorize(ra)),'b')
```

The X axis is population density and the Y axis is actual rate of population growth per individual. Label the graph, move it to a Word document, and write a figure legend that describes a population (real or imaginary) with each theta value.

The theta logistic equation is developed by substituting this modified description of r_a into the classic logistic equation (Equations (6.4) to (6.5))

$$\frac{dN}{dt} = r_m N\left[1 - \left(\frac{N}{K}\right)^{\theta}\right] \tag{6.7}$$

which for the computer is `Dn=rm*n*(1-(n/k)^theta)`.

See the effect of theta on the pattern of population growth toward a carrying capacity. First ask the program to solve the differential equation. (Omit the semicolon if you wish to see what the equation looks like.)

```
>>n=simplify(dsolve('Dn=rm*n*(1-(n/k)^theta)','n(0)=no','t'));
```

Then assign the same values to the variables as you used above:

```
>>no=10;
>>rm=0.6;
>>k=100;
>>t=0:1:25;
>>figure
>>hold on
>>theta=1;
>>plot(t,eval(vectorize(n)),'k')
>>theta=2;
>>plot(t,eval(vectorize(n)),'g')
>>theta=3;
>>plot(t,eval(vectorize(n)),'r')
>>theta=0.3;
>>plot(t,eval(vectorize(n)),'b')
```

The X axis is time and the Y axis is population density. Label the graph and move it to a Word document. Expain, in the figure legend, the effects of theta on population growth.

Logistic Population Growth: Discrete Time Model

Most organisms reproduce only during a defined season, and for them a discrete time model, rather than a continuous time model, is appropriate. Discrete growth is described by difference equations rather than by differential equations. The classic version of the discrete time model is

$$N_{t+1} = N_t + R_m N_t \left(\frac{K - N_t}{K} \right) \tag{6.8}$$

Using a theta version of this model allows us to vary the relationship between the actual rate of growth per individual and the population density (as we did for the continuous model described above):

$$N_{t+1} = N_t + R_m N_t \left[1 - \left(\frac{N}{K} \right)^{\theta} \right] \tag{6.9}$$

Recall that when theta is equal to 1, the model assumes a linear relation between rate of growth per individual and population density and is the same as the classic version of the discrete growth model. In our computer language, this theta version becomes `nprime = nt + rm*nt*(1-(nt/k)^theta)`.

Predicting the density of mouse populations

Dust-transmitted pulmonary disease is a deadly viral infection of the lungs that spreads from the Gray-footed Mouse to humans upon inhalation of dust that contains feces from these mice. Not surprisingly, the number of human deaths per year from this disease is correlated with the abundance of the mouse. For this reason, physicians have asked ecologists to estimate future abundances of the Gray-footed Mouse.

The Gray-footed Mouse reproduces once a year, in the spring, and so discrete time models are appropriate. It is a food generalist, feeding on seeds and insects, and nests in underground burrows where the soil is deep and pliant. Each female forages within a home range, used exclusively by her, that surrounds the burrow. The carrying capacity of the population (K) is set by the density of nesting sites within the area.

You will use this model of growth several times, and so a stored program will save you time. Write (in the program window) the following instructions for the discrete logistic model with theta:

```
function n=logistic(rm,k,theta,no,runlen)
n=[no];
for t=1:runlen
  nprime=n(t)+rm*n(t)*(1-(n(t)/k)^theta);
  if nprime<0
    nprime=0;
  end
  n=[n nprime];
end
```

Store this program as *logistic* on your floppy disk by scrolling down "file" to "save as." Now, try some values of R_m to see the effect on population growth. Keep all other values constant. Begin with an R_m (symbolized `rm`) value of 1.1 and increase it gradually to see the effect. First, indicate that you will be working from your floppy disk and then construct a graph by typing

```
>>cd a:
>>figure
>>hold on
>>k=100; theta=1; runlen=15; no=80;
>>rm=1.1;
>>n=logistic(rm,k,theta,no,runlen);
>>plot(0:runlen,n,'r')
```

Explore a range of realistic values of R_m for different populations of the Gray-footed Mouse (between 1.0 and 3.0) by entering the new R value and then retyping the last two commands. Use different colors or symbols for the curves (see page 159). Adjust the Y axis to extend between 0 and 150 (by scrolling down "edit" to "axes properties"). Label the graph, move it to a Word document, and write a figure legend. How well can you predict population density? What is the effect of the maximum rate of population growth per individual (R_m) on your accuracy of prediction? Notice that under certain conditions the density of mice is difficult to predict, *even though there are no random elements in the equation – all the parameters are exactly specified.*

Bifurcation diagrams: from stable equilibrium to chaos

The effect of R_m on predictability of N can be examined more thoroughly by constructing a bifurcation diagram. This form of diagram illustrates how the location and stability of solutions to an equation depend on one of its parameters. Here, the solution refers to the population density (N), as predicted by the discrete logistic equation, and the parameter is the maximum rate of population growth per individual (R_m).

To construct the bifurcation diagram, write a program that applies the discrete logistic model for 500 time steps. The first step sets up a loop, with R_m varying from 1.50 to 3 in steps of 0.01. Each time the loop is carried out, the new population is returned as the vector n, which is used to generate the next year's population (using the program *logistic* described above). Each simulation begins with a population density of $k + 1$. Here we use a carrying capacity of 100, and so the initial population (`no`) consists of 101 individuals. Theta is set at 1. The command `r*ones(1,501)` gives a vector of 500 numbers, each of which is a population density. The command `plot(rx(401:501), n(401:501),'.')` tells the computer to plot only the last 100 of the population sizes on the Y axis and the R_m values on the X axis. Enter these commands by typing

```
>>figure
>>hold on
>>for rm=1.5:0.01:3
n=logistic(rm,100,1,101,500);
rx = rm*ones(1,501);
plot(rx(401:501),n(401:501),'.')
end
```

(You will have to wait for the computer to compute all these points.) Label the graph with a title (e.g., *Discrete Logistic Model: Relation Between Maximum Rate of Growth and Density*), an X axis (*Maximum Rate of Population Growth* (*R*)), and a Y axis (*Population Density*). The graph shows that for small values of R_m, the population always reaches a stable equilibrium at its carrying capacity; density is completely predictable. Between R_m values of about 2.0 and 2.4, the population oscillates between two densities, one above and the other below the carrying capacity. At higher values of R_m, the population oscillates among four, eight, and then even more densities. When R_m is 3, the population oscillates among so many densities that it is impossible to predict future densities. This type of pattern is known as chaos; it occurs here because a deterministic equation (with no random elements) generates so many possible solutions that the outcome appears random.

Transfer the graph to a Word document and give it a figure legend. What can you tell the physicians about your ability to predict the population density of the Gray-footed Mouse when R_m is less than 2? Equal to 2.2? Equal to 2.5? Equal to 2.7? Equal to 3?

Effect of initial density

Suppose now that you want to study the effects of a vaccine against dust-transmitted pulmonary disease on the population density of the Gray-footed Mouse. You divide (at random) a population into two subpopulations: an experimental one, which will receive the vaccine, and a control, which will not receive the vaccine. Is this a valid experiment if the population growth rate is high? In other words, can you assume that the only difference in future densities between the subpopulations is due to the vaccine? To find out, see what happens to the density of two identical subpopulations growing at an R_m of 3. (Actually, the program will not work if they are absolutely identical. Thus, if your carrying capacity is 100 individuals, then start one population with 99 individuals and the other with 101 individuals.) Run them for 25 years.

```
>>n1=logistic(3,100,1,101,25);
>>n2=logistic(3,100,1,99,25);
>>figure
>>hold on
```

Graph the population densities as a function of time with

```
>>plot(0:25,n1(1:26),'r')
>>plot(0:25,n2(1:26),'b')
```

Label the graph, transfer it to a Word document, and write a figure legend. Explain why populations with high maximum rates of growth should not be used in controlled experiments. This graph also shows another feature of chaos: initial conditions (here, initial densities) can generate very different outcomes.

EXERCISE 7

Interspecific Competition and Coexistence

Interspecific competition is competition between two species for a resource that is limited in supply. The classic model was developed by Vito Volterra, an Italian mathematician, in 1926. Later, in 1932, Alfred Lotka, an American population biologist, developed a way of analyzing the model in graphical form. The model consists of the differental equation for logistic population growth (continuous time) plus a term that describes the negative effect of one species on the other. Here, we will develop the competition equations from the theta version of the logistic equation, because it is more versatile than the classic logistic equation.

$$\frac{dN}{dt} = r_m N\left[1 - \left(\frac{N}{K}\right)^{\theta}\right] \tag{7.1}$$

The equation for the growth of species 1 when it experiences competition from species 2 is written

$$\frac{dN_1}{dt} = r_{m(1)} N_1\left[1 - \left(\frac{N_1}{K_1}\right)^{\theta_1} - \frac{\alpha_{12}N_2}{K_1}\right] \tag{7.2}$$

where $r_{m(1)}$ is the maximum rate of growth per individual for species 1, α_{12} is the effect of an individual of species 2 on an individual of species 1 as compared with the effect of an individual for species 1 on another individual of species 1 (defined as being equal to 1).

Population Ecology: An Introduction to Computer Simulations. By Ruth Bernstein.

When α_{12} is multiplied by the number of individuals of species 2, it describes the overall effect of species 2 in reducing the maximum density of species 1 below its carrying capacity K. The equation for population growth of species 2 is

$$\frac{dN_2}{dt} = r_{m(2)}N_2\left[1 - \left(\frac{N_2}{K_2}\right)^{\theta_2} - \frac{\alpha_{21}N_1}{K_2}\right] \quad (7.3)$$

These equations, which describe population growth of two competing species, are known as the Lotka–Volterra competition model. In this exercise, we examine (1) the condition of stable equilibrium, (2) the effects of theta, and (3) the effects of an environmental change that alters the carrying capacity of one of the species.

The Niches of Two Competing Species of Perch

Suppose that you are a fisheries biologist planning to develop lakes from abandoned gravel pits. You want to stock the lakes with two species of perch – the Spotted Perch and the Striped Perch. Previous work has shown that the carrying capacities of perch are set by the amount of available food, and stomach analyses reveal that both species eat only minnows (the minnow family: Cyprinidae). You conclude from this work that if their niches overlap, with regard to diet, then this overlap is likely to represent interspecific competition. (It is important to realize that an overlap between niches does not necessarily mean competition between the species. If the resource is not in short supply, then there is no competition.)

Before you spend a lot of time and energy in stocking the gravel pits, you decide to quantify the niches of these two species of perch and to simulate competition using the Lotka–Volterra competition equations. You calculate, with regard to diet, the niche breadth of each species and the niche overlap between species. Niches can be quantified in several ways. The following methods were developed, in 1967, by the American population ecologists Robert MacArthur and Richard Levins.

The equation for niche breadth is

$$\text{niche breadth} = \frac{1}{\sum_{k=1}^{n} P_{ik}^2} \quad (7.4)$$

where i is the species under consideration (1 for the Spotted Perch and 2 for the Striped Perch), k is the category of diet, and P is the proportion of the diet that consists of category k. Here, k is a size category of minnows.

The equations for niche overlap are

$$\alpha_{12} = \frac{\sum P_{1k} P_{2k}}{\sum P_{1k}^2} \tag{7.5}$$

$$\alpha_{21} = \frac{\sum P_{1k} P_{2k}}{\sum P_{2k}^2} \tag{7.6}$$

where P_{1k} is the proportion of the diet of species 1 that consists of food class k and P_{2k} is the proportion of the diet of species 2 that consists of food class k. Here, species 1 is the Spotted Perch, species 2 is the Striped Perch, and the food classes k are size categories of minnows. Using these symbols, you compile the data as shown in Table 7.1. Graph this dietary dimension of the niches for both species. First enter the entire range of minnow sizes:

```
>>x=[0 .1 .3 .5 .7 .9 1.1 1.3 1.5 1.7 1.9 2.1 2.3 2.5 2.7 2.9 3.1];
```

Then enter the proportion of each size class in the diet of each species (i.e., P_{1k} for the Spotted Perch and P_{2k} for the Striped Perch). Be sure that you enter the same number of values for both the X and the Y axes.

```
>>y1=[0 .03 .08 .11 .15 .20 .13 .10 .08 .05 .04 .02 .01 0 0 0 0];
>>y2=[0 0 0 .01 .04 .09 .15 .18 .16 .12 .09 .06 .04 .02 .02 .01 0];
```

Construct the two histograms. First,

```
>>figure
>>bar(x,y1)
```

This histogram illustrates the diet of the Spotted Perch. The X axis is size of minnows (midpoints of ranges, in cm) and the Y axis is proportion of the diet that consists of that size class. Second,

```
>>figure
>>bar(x,y2)
```

This histogram illustrates the diet of the Striped Perch. The X and Y axes are the same as for the Spotted Perch. Note that the Spotted Perch (which is smaller) takes smaller minnows than the Striped Perch, but there is overlap in their diets – both species eat minnows in the size classes from 0.5 through 2.3. Label the graphs, move them to a Microsoft Word document, and write figure legends.

Calculate the niche breadths and overlaps by inserting values from the table into the MacArthur–Levins equations:

$$\text{niche breadth of the Spotted Perch} = \frac{1}{0.1198} = 8.347$$

$$\text{niche breadth of the Striped Perch} = \frac{1}{0.1194} = 8.374$$

$$\text{overlap of Striped Perch on Spotted Perch, } \alpha_{12} = \frac{0.0865}{0.1198} = 0.722$$

$$\text{overlap of Spotted Perch on Striped Perch, } \alpha_{21} = \frac{0.0865}{0.1194} = 0.724$$

Table 7.1 Diets of the Spotted Perch (species 1) and the Striped Perch (species 2), where P_{1k} is the proportion of the Spotted Perch's diet in minnow size class k and P_{2k} is the proportion of the Striped Perch's diet in minnow size class k.

Length of minnows (midpoint of size class, in cm)	P_{1k}	P^2_{1k}	P_{2k}	P^2_{2k}	$P_{1k}P_{2k}$
0	0	0	0	0	0
0.1	0.03	0.0009	0	0	0
0.3	0.08	0.0064	0	0	0
0.5	0.11	0.0121	0.01	0.0001	0.001
0.7	0.15	0.0225	0.04	0.0016	0.006
0.9	0.20	0.0400	0.09	0.0081	0.018
1.1	0.13	0.0169	0.15	0.0225	0.0195
1.3	0.10	0.0100	0.18	0.0324	0.018
1.5	0.08	0.0064	0.16	0.0256	0.0128
1.7	0.05	0.0025	0.12	0.0144	0.006
1.9	0.04	0.0016	0.09	0.0081	0.0036
2.1	0.02	0.0004	0.06	0.0036	0.0012
2.3	0.01	0.0001	0.04	0.0016	0.0004
2.5	0	0	0.03	0.0009	0
2.7	0	0	0.02	0.0004	0
2.9	0	0	0.01	0.0001	0
3.1	0	0	0	0	0
total		0.1198		0.1194	0.0865

Predicting the Population Densities of the Two Competing Species of Perch

You know, from previous research, that the Spotted Perch has a maximum rate of growth per individual (r_m) of 0.8 and the Striped Perch has a maximum rate of 0.6. (The Spotted Perch has a higher rate because it is smaller, with a shorter developmental time.) You also know that the Spotted Perch defends territories where the bottom substrate consists of larger stones, which it can use as landmarks, but is randomly distributed where the bottom substrate consists only of smaller stones. Thus, for the Spotted Perch in lakes with larger stones, crowding does not occur until the population approaches its carrying capacity (theta=4); in lakes without larger stones, the distribution is approximately random (theta=1.5). The Striped Perch population is always randomly distributed (theta=1.5) within a lake. Before designing the bottom substrate of the lakes, you want to know the effect of larger stones on the densities of these two species of fish. You do this by simulating population growth under the two conditions of substrate.

Begin the simulations by writing a general program for two-species competition. Open the program window and write the program, *lvcomp*, which describes the Lotka–Volterra competition model (theta version) in MATLAB language:

```
function ndot-lvcomp(t,n)
global rm k alpha theta;
ndot(1,1)=rm(1)*n(1)*(1-((n(1)/k(1))^theta(1))-alpha(1,2)*n(2)/k(1));
ndot(2,1)=rm(2)*n(2)*(1-((n(2)/k(2))^theta(2))-alpha(2,1)*n(1)/k(2));
```

Be careful to enter the exact number of parentheses. The parameters in the model are defined as global, which means they can be varied from one simulation to the next. Save the program as *lvcomp* on your floppy disk.

Tell the computer that you will be working from your floppy disk and that the variables are `rm`, `k`, `alpha`, and `theta`:

```
>>cd a:
>>global rm k alpha theta
```

In assigning values to the variables, there are two entries (one for each species) for r_m, theta, and K: the first entry is for species 1 (the Spotted Perch) and the second is for species 2 (the Striped Perch). There are four entries for alpha: the first and last are for intraspecific competition (1 by definition), the second is the effect of species 2 on species 1 ($\alpha_{12} = 0.722$), and the third is the effect of species 1 on species 2 ($\alpha_{21} = 0.724$). For this first simulation, assume that the bottom substrate consists only of smaller stones (i.e., both theta values are equal to 1.5) and that both carrying capacities are equal to 1000. I began my simulation with 100 individuals (no=100) of each species and ran the simulation for 50 years, using the ordinary differential equation solver #23 (ode23) to find solutions to the equations.

```
>>rm=[0.8; 0.6];
>>k=[1000; 1000];
>>alpha=[1 0.722; 0.724 1];
>>theta=[1.5; 1.5];
>>tspan=[0 50];
>>no=[100;100];
>>[t,n]=ode23('lvcomp',tspan,no);
>>figure
>>plot(t,n)
```

The X axis is time, the Y axis is population density, the blue line is species 1, the green line is species 2. Adjust the Y axis to extend from 0 to 1000 (scroll down "edit" and click "axes properties"). Label the graph, indicating the values of the parameters. Move it to a Word document and write a figure legend.

Simulations of the Lotka–Volterra competition equations are usually presented in graphs with the density of species 1 on the X axis and the density of species 2 on the Y axis. To construct such a graph, you need a second program (which incorporates the first one – *lvcomp* – that you stored earlier). This program, called *lvflow*, analyzes the "flow" of densities that occurs when the competition begins at different initial numbers of individuals (from $N_0 = 0.1K$ to $N_0 = 1.2K$ for each species). For each initial condition, it calls ode23 to generate a solution to the Lotka–Volterra equations. The t_stop command tells the program the length of time it should run.

```
function lvflow(t_stop)
  global rm k alpha theta
  figure
  hold on
  for n1=[0.1*k(1) 1.2*k(1)]
    for n2=0.1*k(2):0.1*k(2):1.2*k(2)
      tspan=[0 t_stop];
      no=[n1;n2];
      [t,n]=ode23('lvcomp',tspan,no);
      plot(n(:,1),n(:,2))
      end
    end
    for n2=[0.1*k(2) 1.2*k(2)];
```

```
    for n1=0.2*k(1):0.1*k(1):1.1*k(1)
    tspan=[0 t_stop];
    no=[n1;n2];
    [t,n]=ode23('lvcomp',tspan,no);
    plot(n(:,1),n(:,2))
    end
  end
```

Save this program as *lvflow* on your floppy disk. Note that the program tells the computer to develop a figure and to hold on to it, so you do not need to type these commands. To see this version of the simulation that you carried out above (using a time period of 50), simply type

```
>> lvflow(50)
```

You may have to wait for the computer to give you an answer. (As this program is long, it is likely that you will make a mistake in typing. Thus, if you get an error message, open the program window to *lvflow* and correct any errors that you find. Then save the program and retype the `lvflow` command.) Label the graph. The X axis is the density of the Spotted Perch and the Y axis is the density of the Striped Perch. Each line represents the flow of the two population densities toward the equilibrium. The lines differ in the densities of the two species at which the simulation began. (If you have trouble interpreting the graph, look at the equilibrium densities that you found in the previous graph.) Add a few arrows to indicate the direction in which the population densities flow toward their equilibrium point (click the arrow icon and then move the cursor along the path where you want the arrow, then click somewhere away from the arrow). Move the graph to a Word document and write a figure legend. What happens to the two populations? Do they coexist? At what densities? What would happen if you eliminated half the Spotted Perch?

Now repeat the simulation for lakes with larger rocks, in which the Spotted Perch forms territories and has a theta value of 4.

```
>>theta=[4; 1.5];
>>[t,n]=ode23('lvcomp',tspan,no);
>>figure
>>plot(t,n)
```

Adjust the Y axis to extend from 0 to 1000. Label the graph, move it to a Word document, and write a figure legend. What is the effect of territorial behavior on equilibrium densities of the two species? What do these simulations tell you about the design of lakes?

Consider a situation in which the Striped Perch is aggressive toward the Spotted Perch, making it less efficient in its feeding activities. The alpha values under these conditions are: $\alpha_{12} = 0.900$ and $\alpha_{21} = 0.724$. Return both theta values to 1.5.

```
>>theta=[1.5; 1.5];
>>alpha=[1 0.9; 0.724 1];
>>[t,n]=ode23('lvcomp',tspan,no);
>>figure
>>plot(t,n)
```

Adjust the Y axis to extend from 0 to 1000. Label the graph, transfer it to a Word document, and write a figure legend. What is the effect of the competition coefficients, alpha, on the outcome of the competition?

Lastly, you are concerned that the minnows you stock may grow larger through time, creating more food for the Striped Perch. Simulate this situation by increasing the carrying capacity of the Striped Perch to 1500, leaving that of the Spotted Perch at 1000.

```
>>k=[1000; 1500];
>>[t,n]=ode23('lvcomp',tspan,no);
>>figure
>>plot(t,n)
```

Adjust the Y axis to extend from 0 to 1000. Label the graph, move it to a Word document, and write a figure legend. What is the effect of a change in the carrying capacity of one of the species?

EXERCISE 8

Interspecific Competition and Geographic Distributions

Competing species not only lower the abundances but also restrict the geographic distributions of one another. Competitive replacement is the term for the situation in which the environment for a species is appropriate but the species is absent due to competition from another species. In this exercise, we consider the effects of interspecific competition on population abundances and on the borders of species distributions.

The Black-crowned Sparrow and the Hermit Sparrow are grassland birds that feed mainly on seeds. When you travel eastward within the grassland, along a gradient of increasing rainfall and therefore of increasing grass height, you see at first only the Black-crowned Sparrow, then both sparrows, and then only the Hermit Sparrow. Interspecific competition occurs within this gradient where the distributions of the two species overlap.

Because the Hermit Sparrow is slightly larger than the Black-crowned Sparrow, it consumes more seeds and therefore has a greater competitive effect on the Black-crowned Sparrow than vice versa. The two competition coefficients are $\alpha_{bh} = 0.8$ and $\alpha_{hb} = 0.6$, where subscript b stands for Black-crowned, subscript h stands for Hermit, α_{bh} is the competitive effect of the Hermit on the Black-crowned, and α_{hb} is the competitive effect of the Black-crowned on the Hermit. Each species *in its optimal environment* has a carrying capacity of 200 individuals per 1000 hectares of grassland.

Population Ecology: An Introduction to Computer Simulations. By Ruth Bernstein.

We now use this information to predict the geographic distributions of the two species and the amount of overlap in their distribution along a 1000 kilometer transect from west to east within the grassland. The alpha values remain constant but the carrying capacities change, due to changes in the relative ability of each species to use the resources that are available. At each point, the program applies the Lotka–Volterra competition model to determine the abundance of each species.

Species Abundances in the Absence of Interspecific Competition

Before beginning our analysis of interspecific competition, it is useful to quantify the abundances of each species in the absence of interspecific competition – to develop a *control* for our study of competition. The Black-crowned Sparrow has a carrying capacity of 200 at the western end of our transect (kilometer 0) that decreases to 100 at the eastern end (kilometer 1000). Assuming a straight-line relationship between carrying capacity (K) and distance along the gradient (x), the equation of the relationship is

$$Y = a + bX \tag{8.1}$$

$$K = 200 + \left(\frac{100 - 200}{1000 - 0}\right)X \tag{8.2}$$

Simplifying the equation and identifying the carrying capacity as that of the Black-crowned, K_b, we get

$$K_b = 200 - \frac{X}{10} \tag{8.3}$$

The Hermit Sparrow has a carrying capacity that is 100 at the western end and increases to 200 at the eastern end. Using the same calculations as for the Black-crowned Sparrow above, we get

$$K_h = 100 + \frac{X}{10} \tag{8.4}$$

The variable X is the position along the gradient – the number of kilometers from the western border. For example, the carrying capacity for the Black-crowned Sparrow at kilometer 300 is

$$K_b = 200 - \frac{300}{10} = 170$$

We need to modify these equations slightly to communicate with MATLAB. Note that, for the Black-crowned Sparrow, the following pair of equations is identical:

$$K_b = 200 - \frac{X}{10}$$

$$K_b = 200\left(1 - \frac{X}{2000}\right)$$

Similarly, for the Hermit Sparrow:

$$K_h = 100 + \frac{X}{10}$$

$$K_h = 100\left(1 + \frac{X}{1000}\right)$$

Type these equations in MATLAB form:

```
>>kb='200*(1-x/2000)';
>>kh='100*(1+x/1000)';
```

Develop a graph by typing

```
>>x=0:50:1000;
>>figure
>>hold on
>>plot(x,eval(vectorize(kb)),'b');
>>plot(x,eval(vectorize(kh)),'r');
```

The X axis is the environmental gradient, from kilometer 1 to kilometer 1000. The Y axis shows the expected densities of the Black-crowned Sparrow (blue) and the Hermit Sparrow (red) along the environmental gradient *in the absence of interspecific competition.* This graph serves as your control – the expected abundances if the two species were not competing. Adjust the Y axis to go from 0 to 200. Label the graph, move it to a Microsoft Word document, and write a figure legend.

Effect of Interspecific Competition on Species Borders

Now see what happens to the geographic distributions when the two species compete with each other. Consider first the Black-crowned Sparrow. In going from east to west, when can this sparrow first occur? The population growth equation for the Black-crowned (when competing with the Hermit) is:

$$\frac{dN_b}{dt} = r_b N_b \left[1 - \left(\frac{N_b}{K_b}\right)^{\theta} - \frac{\alpha_{bh} N_h}{K_b}\right] \qquad (8.5)$$

At the eastern border of the Black-crowned Sparrow, its number drops to zero ($N_b = 0$) and the Hermit Sparrow is at its carrying capacity ($N_h = K_h$). The term inside the brackets becomes

$$1 - 0 - \left(\frac{\alpha_{bh} N_h}{K_b}\right)$$

Rearranging, we get

$$\frac{K_b}{K_b} - \frac{\alpha_{bh} N_h}{K_b}$$

or

$$\frac{K_b - \alpha_{bh} N_h}{K_b}$$

The eastern border of the Black-crowned Sparrow is where this term is equal to 0, which occurs when the numerator is 0 (the denominator cannot be 0). Thus, the eastern border is where

$$K_b - \alpha_{bh}N_h = 0$$

Just west of this border, a few Black-crowned Sparrows occur and the Hermit Sparrows do not quite reach their carrying capacity. Thus, the Black-crowned Sparrow can occur wherever

$$\begin{aligned} K_b - \alpha_{bh}K_h &> 0 \\ K_b &> \alpha_{bh}K_h \\ \frac{K_b}{K_h} &> \alpha_{bh} \end{aligned} \tag{8.6}$$

The western border of the Hermit Sparrow is found in the same way. MATLAB does this for us. As we have already entered the equations for carrying capacities along the environmental gradient, all we need to do is give the program the alpha values

```
abh='0.8';
ahb='0.6';
```

and then tell it how to construct the graphs. First, find the eastern border of the Black-crowned Sparrow:

```
>>figure
>>hold on
>>plot(x,eval(vectorize(symop(kb,'/',kh))),'b');
>>plot([0 1000],[eval(abh) eval(abh)],'b');
```

(The `symop` command constructs a complicated formula from a simple one.) The *Y* axis of this graph is the ratio K_b/K_h. Where this ratio is greater than 0.8 (α_{bh}), the Black-crowned Sparrow can occur; where it is less than 0.8, the Black-crowned Sparrow cannot occur. How does interspecific competition affect the borders of the geographic distribution of this sparrow? Adjust the *Y* axis to go from 0 to 2. Find the exact eastern border of the Black-crowned Sparrow (in kilometers along the 1000 kilometer environmental gradient) by typing

```
>>xb=solve(symop(abh,'-',kb,'/',kh))
```

Label the graph, transfer it to a Word document, and write a figure legend.

Now find the western border for the Hermit Sparrow:

```
>>figure
>>hold on
>>plot(x,eval(vectorize(symop(kh,'/',kb))),'r');
>>plot([0 1000],[eval(ahb) eval(ahb)],'r');
```

Adjust the Y axis to go from 0 to 2. The Y axis of this graph is the ratio K_h/K_b. The border is where this ratio is equal to 0.6 (α_{hb}). How does interspecific competition affect the geographic distribution of the Hermit Sparrow? Find the exact border, in terms of kilometers along the gradient, by typing

```
>>xh=solve(symop(ahb,'-',kh,'/',kb))
```

Label the graph, transfer it to a Word document, and write a figure legend.

Effect of Interspecific Competition on Species Abundances

Interspecific competition not only contracts the geographic distribution of a species but also reduces the abundances of the two species wherever they coexist. Construct a graph of the densities of both species along the 1000 kilometers of environmental change. We have already entered the values for when each species occurs alone; now we need to calculate the densities where they coexist. As you can see below, there are many commands and it is easy to make a mistake. (The `subs` command is used to replace one variable with another.) I suggest that you type the commands in the program window and give them a file name. Then, return to the command window and just type that file name. In this way, when you have made a mistake or when you repeat the exercise with different alpha values (below), you do not need to retype all the commands. Thus, insert a floppy disk, open the program window (scroll down "file" to "new, m-file"), and type

```
nb='(kb-abh*kh)/(1-abh*ahb)';
nb=subs(nb,kh,'kh');
nb=subs(nb,kb,'kb');
nb=subs(nb,ahb,'ahb');
nb=subs(nb,abh,'abh');
nh='(kh-ahb*kb)/(1-abh*ahb)';
nh=subs(nh,kh,'kh');
nh=subs(nh,kb,'kb');
nh=subs(nh,ahb,'ahb');
```

```
nh=subs(nh,abh,'abh');
figure
hold on
x=0:10:eval(xh);
plot(x,eval(vectorize(kb)),'b')
x=eval(xh):10:eval(xb);
plot(x,eval(vectorize(nb)),'b')
plot(x,eval(vectorize(nh)),'r')
x=eval(xb):10:1000;
plot(x,eval(vectorize(kh)),'r')
```

Save these commands as *sparrows* on your floppy disk. Return to the command window. Tell the computer you are working from a floppy disk (`cd a:`) and type `sparrows`. Your graph will appear. Both the red line and the blue line should be relatively smooth (with no major gaps). If the graph does not show two curves, one blue and the other red, check your commands for errors. An error may result from mistyping one of the commands in the *sparrows* program or from an error in the `xh=solve(symop...` command or the `xb=solve(symop...` command.

Label the graph, move it to a Word document, and write a figure legend. If you are counting the densities of the Black-crowned Sparrow along a transect from the western to the eastern border of the grassland, how would you know that a competitor had appeared in the transect?

As a final step in the exercise, change one of the alpha values and then reproduce this last graph. (Do not use two identical alpha values.) Do this by inserting the new value in one of the commands you used before (either `ahb='0.6';` or `abh='0.8'`) and then retyping the `solve` command that involves that particular alpha. If you evaluate the Hermit sparrow, for example, change `ahb` from 0.6 to the new value, retype the `xh=solve...` command, save the program, and then type "`sparrows`". The new graph should appear. A negative position on the X axis means that the distribution of the species extends farther west than kilometer 1 of your transect. Label the graph, transfer it to the Word document, and discuss the effect of alpha on the geographic distributions.

EXERCISE 9

Predator–Prey Dynamics: Introduction to the Model

How do a predator and its prey affect the dynamics of each other's population? What characteristics of the two species determine the equilibrium densities and what characteristics allow for a stable equilibrium? The classic predator prey model, developed by the American population biologist Alfred Lotka in 1920 and 1924 and the Italian mathematician Vito Volterra in 1926, assumes that without predation the prey population grows exponentially without limit. This model is no longer useful, as we have found that its predictions are unrealistic. Instead, we use a modification of this model, known as the logistic model, in which the prey has a carrying capacity and its growth is described by the logistic equation.

The logistic model assumes that in the absence of predators, the prey population grows to its carrying capacity. In this exercise, we describe prey growth with the theta version of the logistic equation for continuous time:

$$\frac{dN_1}{dt} = r_m \left[1 - \left(\frac{N_1}{K_1}\right)^{\theta}\right] \tag{9.1}$$

where N_1 is the population density, r_m is the maximum growth rate per individual, K_1 is the carrying capacity of the environment for the prey population in the absence of predators, and theta describes the relation between the actual growth rate per individual (r_a) and population density (see Exercise 6). Predators are incorporated into the equation by subtracting the rate at which individuals die from predation. This rate depends on the ability of the predator to find, catch, and kill its prey. The ability of a predator to find its

Population Ecology: An Introduction to Computer Simulations. By Ruth Bernstein.

prey increases with prey density (N_1). The ability to catch and kill is described by the parameter a, which is the proportion of prey individuals encountered that are caught and killed. Thus, the rate at which predators remove prey from the population is the ability of each predator to find, catch, and kill (aN_1) times the number of predators (N_2). The complete equation describing prey population growth in the presence of predators is

$$\frac{dN_1}{dt} = r_m\left[1 - \left(\frac{N_1}{K_1}\right)^{\theta}\right] - aN_1N_2 \qquad (9.2)$$

Now consider population growth of the predator. The rate at which individuals are added to the population (i.e., the birth rate) depends on the amount of food consumed by the predator population (aN_1N_2) times the efficiency at which this food is converted into more predators. This efficiency, known as the conversion coefficient b, is the result of many factors, such as the energy spent in hunting, the calories contained within each prey individual, the proportion of the consumed prey that is assimilated into the body cells, the energetic cost of mating, and the energetic cost of parental care. It is rarely more than 15 percent; it is lower in endotherms than in ectotherms. Lastly, the model assumes a constant death rate (d) per predator. The complete equation describing the predator population growth is

$$\frac{dN_2}{dt} = baN_1N_2 - dN_2 \qquad (9.3)$$

where b is the efficiency with which the predator converts its food into offspring, aN_1N_2 is its food supply, and d is the death rate per predator.

How do the parameters of these equations affect the equilibrium densities and the stability of the two populations? Equations for equilibrium densities are obtained by setting the differential equations equal to zero. As you can see from the growth equation for the prey (N_1) above, $dN_1/dt = 0$ when

$$r_mN_1\left[1 - \left(\frac{N_1}{K_1}\right)^{\theta}\right] = aN_1N_2$$

which, when simplified, becomes

$$N_2 = \frac{r_m}{a}\left[1 - \left(\frac{N_1}{K_1}\right)^{\theta}\right] \qquad (9.4)$$

Thus, the prey population has zero growth when the predator population is equal to the term on the right above. For the predator, $dN_2/dt = 0$ when

$$b(aN_1)N_2 = dN_2$$

Simplifying the equality, we get

$$N_1 = \frac{d}{ba} \tag{9.5}$$

Thus, the predator population has zero growth when the density of the prey population is equal to the term on the right.

The following four exercises explore the effects of predator efficiency, social behavior, carrying capacity, and human harvesting on the equilibrium densities and population stabilities of a predator and its prey.

EXERCISE 10

Predator–Prey Dynamics: Effect of Predator Efficiency

This exercise analyzes the dynamics of a population of swallows and the hawks that prey upon it. Read Exericise 9 before you begin your analysis. The Ruby Swallow is an insectivorous bird that nests in tree holes. The carrying capacity of the population is set by the density of trees that can be excavated for nest holes. The population density of the Grayish Hawk is determined by the rate at which it can find, catch, and kill the swallows. The ability to find swallows depends on the density of swallows (N_1). The ability to catch and kill swallows, known as the predator efficiency a, is a coevolved trait reflecting adaptations of the hawks for catching and killing the swallows and adaptations of the swallows for avoiding being caught and killed. (Two species coevolve when the evolution of one influences the evolution of the other, and vice versa.)

Graphical Analysis of Consumption Rate

Assume that the efficiency of hawks at catching swallows (the parameter a) is a constant for each environmental setting. State the variables used in this exercise and then graph the relationship between consumption by each hawk and density of swallows by typing

Population Ecology: An Introduction to Computer Simulations. By Ruth Bernstein.

```
>>global r k theta a b d
>>consumption='a*n1';
>>a=0.01;
>>n1=0:1:1000;
>>figure
>>hold on
>>plot(n1,eval(consumption),'c')
>>a=0.02;
>>plot(n1,eval(consumption),'r')
>>a=0.04;
>>plot(n1,eval(consumption),'g')
```

The X axis is density of swallows, the Y axis is number of swallows that a hawk could consume per unit time, and the slope of each line is a. This relationship between predator consumption and prey density is known as the functional response of the predator (in contrast with its numerical response, in which predator numbers increase with prey density). Where $a = 0.01$, each hawk captures and consumes 1 out of every 100 swallows that it encounters; where $a = 0.02$, each hawk captures and consumes 2 out of every 100 swallows that it encounters, and where $a = 0.04$ it consumes 4 out of every 100. The consumption rate per hawk per year is described by the equation for a straight line: $Y = 0 + aN_1$.

Effect of Predator Efficiency on Equilibrium Densities

The effect of a on the equilibrium densities can be analyzed by a graph that shows the rate at which the prey population adds new individuals to its population (`dn1`) and the points of zero population growth for predators of different efficiencies. To construct such a graph, first enter the equation for logistic growth in the prey population (N_1). Then assign values to the variables:

```
>>dn1='r*n1*(1-(n1/k)^theta)';
>>r=1.0; k=1000; theta=1;
>>n1=[0:1:1000];
>>figure
>>hold on
>>plot(n1,eval(vectorize(dn1)),'k')
```

The X axis is the density of the swallow population and the Y axis is the rate at which new swallows are added to the population each year. In order for the prey population to remain at a constant density, the hawks would need to kill swallows at the same rate as they are added to the population, a rate that varies with swallow density. The curve peaks at $\frac{1}{2}K$, which is always the case when theta = 1, because at that density the population is growing the fastest – at lower densities, there are too few individuals; at higher densities, the rate of growth per individual is depressed by crowding.

Next add to the graph the points at which a predator has zero population growth (see Equation 9.5). As this equation depends on three constants, simply assign values to the variables d, b, and a and calculate $N_1 = d/ab$. I assigned a death rate of 0.3, meaning that 30 percent of the hawk population dies each year, a conversion coefficient of 0.03, meaning that 3 percent of the swallow material consumed by a hawk becomes new adult hawks, and a hawk efficiency of 0.02, meaning that a hawk captures and kills 2 out of every 100 swallows that it encounters. The equilibrium densities of hawks resulting from these values occur when

$$N_1 = \frac{0.3}{(0.02)(0.03)} = 500$$

Place this vertical line on the graph by first defining the Y axis to match the one already on the graph and then giving the `plot` command.

```
>>y=[0:1:300];
>>plot(500,y,'r')
```

The rate at which the hawks can remove swallows (the Y axis) and still maintain a constant density of swallows is the point where the vertical line crosses the curved line.

See what happens in an environmental setting where the hawks are more efficient and a setting where the hawks are less efficient than 0.02. If the hawks are twice as efficient, with $a = 0.04$, then the vertical line representing zero population growth of the predator is at

$$N_1 = \frac{0.3}{(0.04)(0.03)} = 250$$

that is,

```
>>plot(250,y,'g')
```

If the hawks are less efficient, with an a of 0.011, then the line is at $N_1 = 909$:

```
>>plot(909,y,'c')
```

Label this graph, move it to a Microsoft Word document, and write a figure legend. What is the effect of *a* on the sustainable rate of predation? Discuss this effect in terms of the coevolution of the two species. Describe environmental settings that would increase and decrease the hawk's ability to catch and kill swallows.

Effect of Predator Efficiency on Population Stability

Analyze the dynamics of the relation between the Ruby Swallow and the Grayish Hawk. Begin by writing a program. Open the program window (scroll down "file" to "new, m-file") and write the following description of predator–prey dynamics:

```
function ndot=pred(t,n)
global r a b d k theta;
ndot(1,1)=r*n(1)*(1-(n(1)/k)^theta)-a*n(1)*n(2);
ndot(2,1)=b*a*n(1)*n(2)-d*n(2);
```

Save as *pred* on a floppy disk.

To simulate the interaction, first define the parameters. I have chosen a maximum rate of growth (r_m) of 1.0 for these swallows, a carrying capacity (K) of 1000, a theta value of 1, an a of 0.02 (2 in 100 encounters result in a kill), a b of 0.03 (only 3 percent of the consumed swallows becomes new hawks), and a d of 0.15 (each year, 15 percent of the hawk population dies). Enter these values and decide on a time period for the simulation (I chose 100 years) and initial densities for the two populations (I chose 200 swallows and 20 hawks). Tell the computer to solve the equations, using the ordinary differential equation solver # 45 (which uses larger steps and returns fewer data points than # 23), and prepare a graph showing the densities over the time period.

```
>>cd a:
>>r=1; k=1000; theta=1; a=0.02; b=0.03; d=0.3;
>>tspan=[0 100];
>>no=[30;10];
>>[t,n]=ode45('pred',tspan,no);
>>figure
>>plot(t,n)
```

The X axis is time (in years) and the Y axis is population density. Now see what happens when a is larger and smaller. Try $a = 0.01$, 0.1, and 0.2. Simply type in the new a value and then retype the commands beginning with `[t,n]=...`. Label the graphs, move them to a Word document, and write figure legends. What is the effect of predator efficiency on the two equilibrium densities? On stability?

The population dynamics of a predator and its prey are usually graphed in a different way – with prey density on the X axis and predator density on the Y axis. Prepare such a graph by first entering the two differential equations:

```
>>dn1='r*n1*(1-(n1/k)^theta)-a*n1*n2';
>>dn2='b*a*n1*n2-d*n2';
```

In the language of mathematics, the equilibrium density of population 1 (the swallows) is called $\hat{N}_1$, in which the symbol above the N is pronounced "hat." In the language of MATLAB, the equilibrium density of population 1 is written `n1hat` and, similarly, the equilibrium density for population 2 (the hawks) is written `n2hat`. Tell the computer to solve the two differential equations:

```
>>[n1hat,n2hat]=solve(dn1,dn2,'n1,n2');
>>n1hat=simple(n1hat)
>>n2hat=simple(n2hat)
```

MATLAB responds with three solutions for each population, presented in three rows and one column. The equations in the third rows are the ones of interest here and will be used to find population densities; they are called into play by typing "`3,1`", for third row and first column.

Begin your simulations by assigning values to all the variables, except predator efficiency, and then setting up a graph.

```
>>r=1; k=1000; theta=1; b=0.03; d=0.3;
>>figure
>>hold on
```

Now, to save time, save a series of commands that you will use several times. Open the program window and type:

```
a=0.02;
n1hat_n=eval(sym(n1hat,3,1));
n2hat_n=eval(sym(n2hat,3,1));
plot(n1hat_n,n2hat_n,'*')
tspan=[0 300];
no=[1.5*n1hat_n;1.5*n2hat_n];
```

```
[t,n]=ode45('pred',tspan,no);
plot(n(:,1),n(:,2),'b')
```

These commands tell the computer to evaluate and simplify the equations for equilibrium densities using the assigned values, and then to plot these densities with an asterisk on the graph. They also tell the computer to begin the simulations at 1.5 times these equilibrium densities, in order to follow the changes in densities through time. Save this program on your floppy disk, giving it the title of *hawk*. To run a simulation, simply type "hawk":

```
>>hawk
```

On this graph, the X axis is the number of swallows; the Y axis is the number of hawks. To understand the graph, compare it with your first figures, in which you plotted numbers over time. Add arrows to the spirals, to show how the two populations spiral inward toward a stable equilibrium. Do this by clicking first on the right-slanting arrow icon on the toolbar and then clicking and dragging where you want to place the arrow.

Explore the effect of predator efficiency on stability of the interaction. Develop three hunting scenarios that would give rise to different predator efficiencies. Simulate these interactions (use a values somewhere between 0.015 and 0.2) to add to the above graph. For each new simulation, simply open the program *hawk*, enter a different a value and color, and save the revised program. When you return to the command window and type "hawk," the results will appear on the figure.

Label the graph, move it to your Word document, and write a figure legend. Describe how you changed the environment to alter the predator efficiency. What happens to the equilibrium densities and to the stability (i.e., amplitude of the oscillations) as the hawk becomes more efficient (or the swallow becomes easier to catch and kill)? Explain these results in terms of coevolution of the hawk and the swallow.

EXERCISE 11

Predator–Prey Dynamics: Effects of Social Behavior

The basic predator–prey model described in Exercise 9 contains no parameters that address the effects of social behavior by either the predator or its prey. In this exercise you will explore the roles of spatial distribution of the prey population and cooperative behaviors among predators in determining the equilibrium densities and stability of the two interacting populations.

Spatial Distribution of the Prey Population

The distribution of individuals within the habitat is described by theta in the logistic equation. Recall from Exercise 6 that when theta is equal to 1, the actual rate of growth per individual (r_a) decreases as a straight line with increases in population density – each additional individual in the population depresses r_a by the same absolute amount. A theta value greater than 1 indicates that the effects of crowding are greater at higher densities, which would be realistic for populations consisting of widely spaced individuals, such as territorial animals. A theta value less than 1 indicates that the effects of crowding are greater at lower densities than at higher densities, which would be realistic for populations in which individuals are clustered at lower densities, when environmental conditions are poor.

Population Ecology: An Introduction to Computer Simulations. By Ruth Bernstein.

State the variables used in this exercise and then construct a graph showing the rate at which new individuals are added to prey populations with theta values of 1, 5, and 0.3. Assume that all three populations have the same maximum rate of growth and carrying capacity.

```
>>global r k theta a b d
>>dn1='r*n1*(1-(n1/k)^theta)';
>>r=1.3; k=400;
>>theta=1;
>>n1=[0:1:400];
>>figure
>>hold on
>>plot(n1,eval(vectorize(dn1)),'k')
>>theta=5;
>>plot(n1,eval(vectorize(dn1)),'m')
>>theta=0.3;
>>plot(n1,eval(vectorize(dn1)),'g')
```

Add to this graph a line of zero population growth for the predator. Assume that the values of the variables are the same regardless of the particular prey population that it is hunting. These variables are $a = 0.075$ (the predator catches 7.5 of every 100 prey individuals that it encounters), $b = 0.02$ (2 percent of the prey material consumed becomes new predators), and $d = 0.3$ (each year, 30 percent of the predator population dies). As described in Exercise 9, the prey density at which the predator has zero population growth is $N_1 = d/ab$, which, for this predator, is 200.

```
>>y=[0:1:350];
>>plot(200,y,'r')
```

On this graph, the X axis is the density of prey, the Y axis is the rate at which the prey population adds individuals to the population, the curved lines are points of zero population growth for the prey population (three populations, each with a different theta value), given that the predator removes prey as rapidly as they are added to the population. The vertical red line is formed of all points in which the predator has zero population growth. Where a curved line intersects the vertical line, the two populations are in equilibrium. Label the graph, move it to a Microsoft Word document, and write a figure legend. Which of the three prey populations should the predator (given its present a, b, and d values) hunt in order to maximize available food? What would have to happen to a in order for a predator population to switch from one prey population to another? How would natural selection work to generate such a change?

Now see what happens to predator–prey dynamics, with different theta values, as the two populations interact through time. Begin by writing a program for interactions between the predator and its prey. Open the program window (by scrolling down the "file" icon to "new, m-file") and type the following *pred* program (if you have not already done so in Exercise 10):

```
function ndot=pred(t,n)
global r k theta a b d;
ndot(1,1)=r*n(1)*(1-(n(1)/k)^theta)-a*n(1)*n(2);
ndot(2,1)=b*a*n(1)*n(2)-d*n(2);
```

Save as *pred* on your floppy disk.

Consider a population of Rusty Foxes that feed on four species of mouse: the Sagebrush Mouse, with a theta of 1; the Grassland Mouse, with a theta of 4; the Willow Mouse, with a theta of 0.3; and the Pine Mouse, with a theta of 0.1. These theta values result from the spatial distribution of burrowing sites, which varies from one habitat to another. All four species of mice have a maximum rate of population growth (r_m) of 1 and a carrying capacity (K) of 800. Use the Sagebrush Mouse for the first simulation (`theta=1`). The fox has an efficiency of 0.1, a conversion coefficient of 0.03, and a death rate of 0.15 per year. Begin the simulation with 100 mice and 25 foxes and use the ordinary differential equation solver # 45 (which uses larger steps and returns fewer data points than # 23). Open the command window and enter

```
>>cd a:
>>r=1; k=800; theta=1; a=0.1; b=0.03; d=0.15;
>>tspan=[0 100];
>>no=[100; 25];
>>[t,n]=ode45('pred',tspan,no);
>>figure
>>plot(t,n)
```

The X axis is time and the Y axis is population density; the blue curve is the Sagebrush Mouse population and the green curve is the fox population. Label the graph (indicating the theta value). Now see what happens to the other species of mice, which have different values of theta.

```
>>theta=4;
>>[t,n]=ode45('pred',tspan,no);
>>figure
>>plot(t,n)
>>theta=0.3;
```

```
>>[t,n]=ode45('pred',tspan,no);
>>figure
>>plot(t,n)
>>theta=0.1;
>> [t,n]=ode45('pred',tspan,no);
>>figure
>>plot(t,n)
```

Label the three figures, move them to a Word document, and write figure legends. What is the effect of theta on stability of the populations?

Predator–prey dynamics are most often presented on graphs in which the X axis is prey density and the Y axis is predator density. See what your fox–mouse interactions look like, with different values of theta, on such a graph. Begin by entering the two equations for population growth (see Exercise 9), in which `n1` is the mouse population and `n2` is the fox population, and then have MATLAB solve the equations for equilibrium densities.

```
>>dn1='r*n1*(1-(n1/k)^theta)-a*n1*n2';
>> dn2='b*a*n1*n2-d*n2';
>>[n1hat,n2hat]=solve(dn1,dn2,'n1,n2');
>>n1hat=simple(n1hat)
>>n2hat=simple(n2hat)
```

Each equation has three solutions. The one we want to use is in the third row and first column (actually, there is only one column). This solution is called up later by typing "`3,1`." Assign values to all the variables except `theta`.

```
>>r=1; k=800; a=0.1; b=0.03; d=0.15;
>>figure
>>hold on
```

Now, so you do not have to keep typing the same commands repeatedly, write a short program telling the computer how to arrange the graph. Open the program window and type

```
theta=1;
n1hat_n=eval(sym(n1hat,3,1));
n2hat_n=eval(sym(n2hat,3,1));
plot(n1hat_n,n2hat_n,'*')
no=[1.5*n1hat_n;1.5*n2hat_n];
tspan=[0 200];
[t,n]=ode45('pred',tspan,no);
plot(n(:,1),n(:,2),'k');
```

Store these commands as *fox* on your floppy disk. Return to the command window and type

```
>>fox
```

This command generates a graph with the density of mice on the X axis and the density of foxes on the Y axis. If you have trouble interpreting it, look at your graph of density against time for theta = 1.

Now, explore the effect of spatial distribution of the mice, using the same values of theta as above. Open the *fox* program, change the value of theta and the color, and then save this revised program. Return to the command window and type "fox." Keep a record of the colors or line styles used for each value of theta. Place arrows on the curved lines (by clicking first the right-slanting arrow icon on the toolbar and then clicking and dragging where you want to place the arrow), showing the direction in which the two densities flow. Label the graph, move it to a Word document, and write a figure legend. What is the effect of theta on equilibrium density of the predator? Of the prey? What is the effect on stability of the populations?

Cooperation among the Predators

The foraging efficiency of a predator may reflect its social behavior. Consider here a population of coyotes that cooperate with one another while hunting rabbits. Without cooperation, the efficiency of a solitary coyote is 0.1 (it catches and kills 1 out of every 10 rabbits that it encounters). With cooperation, the coyote efficiency increases as the number of coyotes increases. Graph this relationship:

```
>>efficiency='0.1+c*n2';
>>n2= 0:1:20;
>>figure
>>hold on
>>c=0.001;
>>plot(n2,eval(efficiency),'r')
>>c=0.005;
>>plot(n2,eval(efficiency),'k')
>>c=0.01;
>>plot(n2,eval(efficiency),'b')
```

The X axis is number of predators and the Y axis is predator efficiency – the probability that prey encountered will be caught and killed. The slope of each line, symbolized c, reflects the advantage to the predators of cooperating in the hunt. Thus, the equation for predator efficiency with cooperation is

```
a = 0.1 + c*n(2)
```
(11.1)

where `n(2)` is the number of predators.

Explore the effect of cooperation on the equilibrium densities and stability of the coyote–rabbit interaction. The rabbit population has a maximum rate of growth (r_m) of 1.5, a carrying capacity of 100, and a theta value of 1. The coyotes have a conversion coefficient of 0.03 (3 percent of the consumed rabbit material becomes new adult coyotes) and a death rate of 0.15 (each year, 15 percent of the population dies). Begin the simulations by entering all values of the variables except predator efficiency (a).

```
>>r=1.5; k=100; theta=1; b=0.03; d=0.15;
>>figure
>>hold on
```

Now enter the values of a. Assuming you already have the program *pred* and the program *fox* stored on your disk, begin by opening the *fox* program window and changing the first command from `theta = 0.1` to `a = 0.1`. Save this revised program as *coyote* on your floppy disk. Return to the command window and type

```
>>coyote
```

The X axis is the population density of rabbits and the Y axis is the population density of coyotes. This particular simulation assumes no social behavior. Now add cooperation by substituting Equation (11.1) for `a` in the program. Begin with `a = 0.1+0.01*n(2)`. Choose another color. Then try several different slope values, ranging from 0.01 to 0.05 – the greater the slope, the more each coyote gains from cooperative hunting. Keep a record of the slope values you chose and the colors you used for each simulation.

Label the graph, move it to a Word document, and give it a figure legend. What is the effect of cooperative hunting on the equilibrium densities of the two species? What is the effect on the stability of the interaction? What do these results suggest about the evolution of cooperative behavior in predators?

Lastly see what happens if the solitary coyote is much less efficient than in the simulation performed above. Set up a new graph (figure hold on) and then modify the equation for efficiency to:

```
a = 0.06 + c*n(2)
```

First see the dynamics of a coyote hunting by itself ($a = 0.06$). Then add cooperative hunting, using c values ranging from 0.0005 to 0.003. Label the graph, move it to a Word document, and give it a figure legend. What is the effect of cooperative hunting on the equilibrium densities of the two species? Comparing the two simulations with different predator efficiencies for solitary coyotes, what can you conclude about the evolution of cooperative behavior in predators?

EXERCISE 12

Predator–Prey Dynamics: Effects of Carrying Capacity and Satiation

In this exercise, you will analyze the effect of carrying capacity with (1) a predator that has no limit to the rate at which it consumes prey, and (2) a predator that shows a limit to its rate of consumption. Read Exercise 9 before simulating these predator–prey interactions.

Consider a situation in which you are a conservation biologist concerned about the low densities of a unique type of lynx on an island. To increase the density of lynx, you plan to increase the density of hares, by converting some of the dense forest to meadow, so there will be more grasses and shrub willows for the hares to eat. Before you spend the time and energy involved in altering the landscape of the island, you decide to run computer simulations of your management plan.

Studies of the two species reveal that there are now 140 hares and 20 lynx on the island. The hare has a maximum r of 0.9, a theta value of 1 (assumes a linear relation between the actual rate of growth per individual and the population density; see Exercise 6), and a carrying capacity (K) of 200. The lynx has a conversion coefficient (b) of 0.02 (only 2 percent of the consumed hare tissue becomes more lynx individuals) and a d of 0.1 (each year, 10 percent of the lynx population dies). You assume that the lynx can consume only a certain number of hares within a time period, but are aware that it may store some of the hare that it catches but does not eat right away. Thus, you decide to simulate both situations – without predator satiation and with predator satiation.

Population Ecology: An Introduction to Computer Simulations. By Ruth Bernstein.

Predator Consumption Rates

The basic predator–prey model, presented in Exercise 9, assumes that the consumption rate per predator increases with prey density in a linear fashion – there is no upper limit to how much a predator can eat (or store). State the variables used in this exercise and then graph the relationship by typing

```
>>global r k theta a b d c
>>consumption='a*n1';
>>n1=0:1:200;
>>a=0.05;
>>figure
>>hold on
>>plot(n1,eval(consumption),'k')
```

The X axis is prey density and the Y axis is number of prey consumed per predator per unit time. This relationship is known as the functional response of the predator, in contrast with a numerical response, in which the predator population grows in response to more prey.

Now consider the situation in which, beyond a certain prey density, the predators have more food than they can use. This situation, known as predator satiation, occurs when the predator is so full that it has no drive to hunt for more prey or when it runs out of hunting time – there is an upper limit to the number of prey that a predator can find, catch, and kill within a foraging period. To model this situation, you need to write a term for the functional response so that predator satiation can be incorporated into the predator–prey model. The relation will be nonlinear, because the predator consumes less and less as it approaches satiation. Several equations describe this relationship. We use here the following:

$$c(1 - \mathrm{e}^{-aN_1/c}) \tag{12.1}$$

In this equation, c is the rate of prey consumption at which the predator becomes satiated, e stands for the base of the natural log, a is the predator efficiency, and N_1 is the density of prey.

See what the relationship between prey consumption and prey density looks like on a graph, using the same value for a as you did for the linear relationship ($a = 0.05$). Choose a value of c (I chose satiation at 10 hares per lynx).

```
>>consumption='c*(1-exp(-a*n1/c))';
>>c=10;
>>plot(n1,eval(consumption),'r')
```

This curve will appear on the same graph as the curve showing the linear relation between consumption and prey density (shown in black). Both curves are different forms of functional response of a predator – how an individual responds to changes in food density. Label the graph, move it to a Microsoft Word document, and write a figure legend.

Now consider what happens to the dynamics of the predator–prey interaction when predator satiation is incorporated into the model. First, write a program *pred* for population dynamics without predator satiation (if you have not already done so in Exercise 10 or 11). Open the program window (by scrolling down "file" to "new, m-file") and type

```
function ndot=pred(t,n)
global r a b d k theta;
ndot(1,1)=r*n(1)*(1-(n(1)/k)^theta)-a*n(1)*n(2);
ndot(2,1)=b*a*n(1)*n(2)-d*n(2);
```

Save as *pred* on your floppy disk. Now write a program for population dynamics with predator satiation.

```
function ndot=predsat(t,n)
global r k theta a c b d;
ndot(1,1)=r*n(1)*(1-(n(1)/k)^theta)-c*(1-exp(-a*n(1)/c))*n(2);
ndot(2,1)=b*c*(1-exp(-a*n(1)/c))*n(2)-d*n(2);
```

Save this program as *predsat* (for predator satiation) on a floppy disk.

Increases in the Prey's Carrying Capacity with a Predator that Does Not Become Satiated

In the command window, tell the computer to work from your floppy disk and enter the equations for density changes in the prey and in the predator. Have the program solve for the two equilibrium densities.

```
>>cd a:
>>dn1='r*n1*(1-(n1/k)^theta)-a*n1*n2';
```

```
>>dn2='b*a*n1*n2-d*n2';
>>[n1hat,n2hat]=solve(dn1,dn2,'n1,n2');
>>n1hat=simple(n1hat)
>>n2hat=simple(n2hat)
```

Each equation has three solutions. The solution in the third row and first column (actually, there is only one column) is the one used here. It will be called up later by typing "3,1." Assign values to the variables and begin the simulation with 140 hares and 20 lynx, the numbers now on the island. Run the simulation for 200 years, using the ordinary differential equation solver # 45 (which uses larger steps and returns fewer data points than # 23).

```
>>r=0.9; k=200; theta=1; a=0.05; b=0.02; d=0.1;
>>tspan=[0 200];
>>no=[140;20];
>>[t,n]=ode45('pred',tspan,no);
>>figure
>>plot(t,n)
```

The *X* axis is time (in years) and the *Y* axis is population density, with the hares in blue and the lynx in green. Now repeat the simulation with *K* values of 500, 1000, and 2000, reflecting possible changes in the island landscape to increase the abundances of hares. To do this, simple type "k = ..." and then begin the new simulation at the "[t,n]= . . . ." command. Label the graphs, move them to a Word document, and write figure legends.

The population dynamics of a predator and its prey are usually presented with the prey density on the *X* axis and the predator density on the *Y* axis. Prepare such a graph by first entering the following commands in the program window (scroll down "file" to "new, m-file"), so you will not have to retype them for each simulation.

```
k=200;
n1hat_n=eval(sym(n1hat,3,1));
n2hat_n=eval(sym(n2hat,3,1));
tspan = [0 200];
plot(n1hat_n,n2hat_n,'*')
no=[1.5*n1hat_n;1.5*n2hat_n];
[t,n]=ode45('pred',tspan,no);
plot(n(:,1),n(:,2),'k')
```

Save as *nosatiation* on your floppy disk. Then type

```
>>figure
>>hold on
>>nosatiation
```

On this graph, the X axis is hare density and the Y axis is lynx density. If you have trouble interpreting the curve, refer back to your graphs of density through time. Try several different values of K by opening the *nosatiation* program, changing the carrying capacity and the color or line style, and then saving the revised program. Place arrows (by clicking on the right-slanted arrow on the toolbar and then clicking and dragging where you want to place the arrow) on the curved line to show the flow of population densities toward equilibrium (represented by an asterisk). Label the graph, move it to a Word document, and write figure legends. What happens to the equilibrium densities and stability as the carrying capacity of the prey increases? What do these results tell you about your plans to raise the carrying capacity of the hares on the island?

Increases in the Prey's Carrying Capacity with a Predator that Becomes Satiated

Laboratory experiments indicate that a lynx becomes satiated after consuming 10 hares. What effect will this new information have on your management plan? Repeat the above simulations but substitute a consumption rate that accounts for predator satiation.

Enter the two differential equations and have the program find solutions to the densities at which both populations have zero growth:

```
>>dn1='r*n1*(1-(n1/k)^theta)-c*(1-exp(-a*n1/c))*n2';
>>dn2='b*c*(1-exp(-a*n1/c))*n2-d*n2';
>>[n1hat,n2hat] = solve(dn1,dn2,'n1,n2');
>>n1hat=simple(n1hat)
>>n2hat=simple(n2hat)
```

The equations in row 3 and column 1 are the ones to use. They will be called up later by typing "3,1." Assign the same values to the variables as you did above, with the addition of a c value of 10.

```
>>r=0.9; k=200; theta=1; c=10; a=0.05; b=0.02; d=0.1;
>>tspan=[0 500];
>>no=[140; 20];
>>[t,n]=ode45('predsat',tspan,no);
>>figure
>>plot(t,n)
```

The X axis is years and the Y axis is population densities. Repeat the simulation using K values of 400, 600, 1000, and 1200. Do this by entering a different k value just before the "`[t,n]= ....`" command. Label the graphs, move them to a Word document, and write figure legends. What is the effect of more hares on the equilibrium densities and stability? How does this result compare with the effect of more hares without satiation? This result is known as the "paradox of enrichment," in which enrichment refers to an increase in the food supply.

Now explore the effect of carrying capacity on a graph with hare density on the X axis and lynx density on the Y axis. Enter the following commands in the program window, so you will not have to retype them for each simulation (you may want to simply open the *nosatiation* program and change the `pred` command to `predsat`). Begin as before with $k = 200$.

```
k=200;
n1hat_n=eval(sym(n1hat,3,1));
n2hat_n=eval(sym(n2hat,3,1));
tspan=[0 200];
plot(n1hat_n,n2hat_n,'*')
no=[1.5*n1hat_n;1.5*n2hat_n];
[time n]=ode45('predsat',tspan,no);
plot(n(:,1),n(:,2),'k')
```

Save as *satiation* on your floppy disk. Return to the command window and type

```
>>figure
>>hold on
>>satiation
```

Repeat the simulation, by altering K in the *satiation* program, saving it, and then returning to the command window to type "`satiation`." Try several different carrying capacities, changing the color or line style for each so that you can keep a record of the simulations. Label the graph, move it to a Word document, and write a figure legend. What is the effect of carrying capacity on equilibrium densities? On stability? How do your conclusions from this satiation model differ from your conclusions from the nonsatiation model?

EXERCISE 13

Predator–Prey Dynamics: Harvesting a Prey Population

Humans often assume the role of predator. We harvest deer and pheasants for sport, for example, and we harvest salmon and lobsters for profit. The difference between this kind of predation, known as harvesting, and natural predation is that we can set the level of harvesting to optimize the yield while at the same time providing long-term stability of the harvested population. In this exercise you will explore the optimal levels of harvesting and the effect on those levels of the spatial distribution of the harvested population.

The Harvesting Equation

A harvesting model only requires terms that describe the dynamics of the prey population and the level of harvesting. It requires no terms for the dynamics of a predator population, as the rate at which prey are removed from the population is determined by humans. In this exercise, we describe the dynamics of the prey population with the continuous model of logistic growth:

Population Ecology: An Introduction to Computer Simulations. By Ruth Bernstein.

$$\frac{dN}{dt} = r_m N\left[1 - \left(\frac{N}{K}\right)^{\theta}\right] \quad (13.1)$$

For the first part of this exercise, we set theta equal to 1 in order to simplify the mathematics. By setting theta equal to 1, the term can be eliminated from the equation. Thus, the equation for growth of the prey population becomes

$$\frac{dN}{dt} = r_m N\left(1 - \frac{N}{K}\right) \quad (13.2)$$

Adding a term for a constant rate of harvesting (h), we have

$$\frac{dN}{dt} = r_m N\left(1 - \frac{N}{K}\right) - h \quad (13.3)$$

which we will enter in MATLAB as: `dn=r*n*(k-n)/k-h`.

Consider a population of Blue Cod that you plan to harvest. Begin by stating the variables used in this exercise. Then enter the equation for logistic growth with harvesting and the commands for constructing a graph that shows the relation between productivity and density of the cod population in the absence of harvesting. Previous studies have shown that this population of Blue Cod has a maximum rate of population growth (r_m) of 1.2 and a carrying capacity of 1000.

```
>>global r k h theta
>>dn='r*n*(k-n)/k-h';
>>r=1.2; k=1000; h=0;
>>n=[0:10:1000];
>>figure
>>plot(n,eval(vectorize(dn)),'k')
```

On this graph, the X axis is the density of cod, from 0 to K, and the Y axis is the number of cod that are added to the population each year. For long-term stability of the population, these additions to the population are harvested. If you remove 300 individuals per year (i.e., a point on the Y axis), then the population will remain stable at 500 individuals (i.e., the point on the X axis that matches 300 on the Y axis). If you harvest 100 individuals per year, you can maintain a population of either 90 cod or 920 cod – both are points of equilibrium because both are points where the level of harvesting equals a rate at which new individuals are added to the population.

The maximum rate at which this population can be harvested is 300 individuals per year. Note that at this rate of harvesting, the population density is halfway between zero and carrying capacity. This rate is the maximum sustainable yield – maximum because it is equal to the maximum rate at which individuals are added to the population; sustainable because a higher rate would cause a decline in the population. At densities lower than 300, the rate of growth per individual is higher, but there are too few individuals for high productivity; at higher densities, there are many more individuals but the rate of growth per individual is depressed by crowding.

Label the graph, move it to a Microsoft Word document, and write a figure legend.

Calculating the Maximum Sustainable Yield

If you harvest at the maximum sustainable yield, the rate at which you remove cod from the population is equal to the rate at which new individuals are produced when the population is growing most rapidly – when it is halfway between zero and carrying capacity, or $\frac{1}{2}K$ (which is always true when theta equals 1). (Maximum sustainable yield varies with theta, as explored in the final section of this exercise.) The maximum sustainable yield when theta equals 1 is easily calculated by substituting $\frac{1}{2}K$ for N in Equation (13.3):

$$\frac{\mathrm{d}N}{\mathrm{d}t} = r_{\mathrm{m}}\frac{K}{2}\left(1 - \frac{\frac{K}{2}}{K}\right) - h \qquad (13.4)$$

$$\frac{\mathrm{d}N}{\mathrm{d}t} = r_{\mathrm{m}}\left(\frac{K}{4}\right) - h \qquad (13.5)$$

Thus, the maximum yield for a stable population (with theta equal to 1) is when $\mathrm{d}N/\mathrm{d}t = 0$ in the above equation, which occurs when $h = r_{\mathrm{m}}K/4$. For the population of Blue Cod, $r_{\mathrm{m}} = 1.2$ and $K = 1000$, which means that the maximum number of fish that can be harvested is 300, at which point the population is stable at 500 individuals. This density is optimal for harvesting the cod. Note that the calculated maximum rate of harvesting is the same as depicted in the graph.

Harvesting at Less Than the Maximum Sustainable Yield

You may decide not to harvest at the maximum sustainable yield, because of a concern that unexpected variations in r_m or K may lead to overharvesting and a decline in the cod population. A more conservative rate of harvesting, somewhat less than maximum, may be more prudent for a long-term fishing plan. Consider a harvesting rate that keeps the harvested population at 15 percent below the density for maximum harvesting. Enter the equation for calculating the rate at which new individuals are added to the population at each population density:

```
>>ddn=simplify(diff(dn,'n'));
```

Ask the program to solve the equation with regard to cod density (n). The conservative solution, abbreviated *nconserve*, applies a harvesting level that is below the maximum sustainable density. The degree to which the population is maintained at less than optimum for maximum harvesting is symbolized by the word *safety*:

```
>>nconserve=solve(symop(ddn,'-','safety'),'n');
```

Tell the program to solve the equation with regard to harvesting (h):

```
>>hconserve=simplify(solve(subs(dn,nconserve,'n'),'h'));
```

Now have the program calculate the equilibrium density:

```
>>nhat=solve(dn,'n')
```

Surprisingly, the cod population has two equilibrium densities, one at row 1, column 1 and the other at row 2, column 1. The two densities are abbreviated as `nhat1` and `nhat2`. Label the two equilibrium densities and tell MATLAB to rearrange the equations into forms that are more simple:

```
>>nhat1=simple(sym(nhat,1,1));
>>nhat2=simple(sym(nhat,2,1));
```

Construct a graph showing cod productivity as a function of cod density, with lines that indicate this conservative level of harvesting and the two equilibrium densities.

```
>>safety=0.15;
>>n=0:10:1000;
>>hconserve_n=eval(hconserve);
>>nhat1_n=eval(subs(nhat1,hconserve,'h'));
>>nhat2_n=eval(subs(nhat2,hconserve,'h'));
>>figure
```

```
>>hold on
>>plot(n,eval(vectorize(dn)),'k')
>>plot([0 k], [hconserve_n hconserve_n],'r')
>>plot([nhat1_n nhat1_n], [0 hconserve_n], 'b')
>>plot([nhat2_n nhat2_n], [0 hconserve_n], 'g')
```

On this graph, the X axis is density of the cod population, from zero to K, and the Y axis is rate at which new individuals are added to the unharvested fish population. The horizontal red line is a rate of harvesting that maintains a population of cod that is 15 percent less than maximum. The vertical lines show the two densities at which the population is in equilibrium with that level of harvesting: green for the second equilibrium (`nhat2`), which is the density at which you planned to maintain the population, and blue for the first equilibrium (`nhat1`). Label the graph, move it to a Word document, and write a figure legend.

Both equilibrium densities are stable with the given rate of harvesting. Now decide which of the two densities is best for your fishing plan. The density at just over 400 (`nhat2`) would seem best, as the cod population can be maintained at a lower density. Yet, predator–prey theory predicts that equilibrium densities to the left of the production peak (where the slope is positive) are less stable than densities to the right (where the slope is negative) when the population is perturbed from its equilibrium density. You know that at the beginning of each fishing season, the cod population is not exactly at its equilibrium density, due to off-season events that cause deviations in the equilibrium density. For each of the two densities of the cod population, see what happens in response to deviations from the equilibrium (plus 25 individuals and minus 25 individuals) just prior to the fishing season.

Begin the simulations by opening the program window (by scrolling down "file" to "new, m-file") and writing a program

```
function dn=harvest(t,n)
  global r k h;
  if (n>0)
    dn=r*n*(k-n)/k-h;
  else
    dn=-n;
  end
```

Save as *harvest* on your floppy disk.

Now tell the computer to work from the floppy disk and start each population at three different initial numbers: the equilibrium density (`nhat1` and `nhat2`), 25 individuals above each density, and 25 individuals below each density.

```
>>cd a:
>>r=1.2; k=1000; h=hconserve_n;
>>tmax=40;
>>figure
>>hold on
>>tspan=[0 tmax];
>>no=[nhat1_n+25];
>> [t,n]=ode23('harvest',tspan,no);
>>plot(t,n,'b')
>> no=[n1hat_n-25];
>>[t,n]=ode23('harvest',tspan,no);
>>plot(t,n,'b')
>>plot([0 tmax], [nhat1_n nhat1_n],'b')
>>no=[nhat2_n+25];
>>[t,n]=ode23('harvest',tspan,no);
>>plot(t,n,'r')
>>no=[nhat2_n-25]
>>[t,n]=ode23('harvest',tspan,no);
>>plot(t,n,'r')
>>plot([0 tmax], [nhat2_n nhat2_n],'r')
```

This graph shows the effects of a disturbance, during the off season, on the population of harvested Blue Cod. The disturbance produces different initial densities (±25 of the equilibrium density) at the beginning of the fishing season. The X axis is time in years; the Y axis is density of the fish population. Blue lines indicate the first equilibrium density (`nhat1`) and red lines indicate the second equilibrium density (`nhat2`). What is the effect of initial density on long-term stability of the population? Do these results confirm the theory, which predicts that the lower equilibrium density (`nhat2`) is less stable than the higher equilibrium density (`nhat1`)? Why does the population, when started at `nhat1+25`, converge to the `nhat2` density? Hint: follow these changes on your second graph, which shows the production rate of the cod population at this level of harvesting. Label the graph, move it to a Word document, and write a figure legend.

Effect of Theta

In your work so far, you have assumed that theta is equal to 1, which means that the population growth rate per individual (r_a) decreases in a linear fashion with increasing density. This assumption may be unrealistic for a population of Blue Cod. See what happens to long-term stability of the cod population when theta is either greater than 1 (i.e., the fish tend to avoid one another) or theta is less than 1 (i.e., the fish tend to cluster when their densities are low).

Begin by constructing a graph that shows productivity of the cod population as a function of density for different values of theta. (I chose theta values of 1, 3, and 0.7.)

```
>>dn='r*n*(1-(n/k)^theta)';
>>r=1.2; k=1000;
>>theta=1;
>>n=[0:10:1000];
>>figure
>>hold on
>>plot(n,eval(vectorize(dn)),'k')
>>theta=3;
>>plot(n,eval(vectorize(dn)),'r')
>>theta=0.7;
>>plot(n,eval(vectorize(dn)),'g')
```

On this graph, the X axis is cod density and the Y axis is the number of new individuals added to the population each year. Label the graph, move it to a Word document, and write a figure legend. What is the effect of theta on maximum sustainable yield? What is the effect of theta on the population density that yields this maximum sustainable yield?

Write a modification of the *harvest* program in which theta is added to the logistic equation of prey growth:

```
function dn=harvesttheta(t,n)
  global r k theta h;
  if (n>0)
    dn=r*n*(1-(n/k)^theta)-h;
  else
    dn=-n;
  end
```

Store as *harvesttheta* on your floppy disk.

Construct a graph showing population density through time using different values of theta. Begin with `theta=1` as a form of control. Use the same harvest rate (`h = 242.5` individuals) per year, and initial numbers (`no = 580`) that generated stable populations above.

```
>>cd a:
>>r=1.2; k=1000; h=242.5; theta=1;
>>tmax=[20];
>>tspan=[0 tmax];
```

```
>>no=580;
>>figure
>>hold on
>>[t,n]=ode23('harvesttheta',tspan,no);
>>plot(t,n,'r')
```

The *X* axis is time; the *Y* axis is population density. Now try different values of theta (e.g., `theta = 3`; `theta = 0.7`) to see the effect of the spatial distribution of individuals within the cod population. Do this by typing "`theta = ....`" and then the commands beginning with "`[t,n] = ....`" Use different colors or line styles so you can keep a record of the effects of each theta. Label the graph, move it to a Word document, and write a figure legend. Why does theta have this effect on population densities and stability?

EXERCISE 14

Optimal Foraging: Searching Predators that Minimize Time

The population density of most terrestrial predators is limited by their food supply. This means that individuals who obtain more than their "share" of the food supply leave more offspring than individuals who are less adept at finding, catching, and killing their prey. Over the generations, the foraging behavior of the population should become more and more efficient with regard to time spent hunting, energy expended in hunting, or both. This area of study is called optimal foraging theory.

There are two general categories of predators: searching and sit-and-wait. A searching predator moves through its habitat and captures the prey items that it wants as it finds them. A sit-and-wait predator positions itself at a vantage point and waits for an unsuspecting prey item to wander near. In this exercise, you explore the behavior of a searching predator whose optimal foraging strategy is to acquire the most food in the least amount of time.

Consider a shrew, which moves across the forest floor searching for insects, worms, snails, and other small invertebrates to eat. Its foraging activity makes it more vulnerable to its own predators – mainly owls. The more time that a shrew spends foraging, the more likely it will be killed. A successful shrew (one who leaves more offspring) makes decisions that give it the most food in the least amount of time. It does this by continually deciding whether to stop and eat a particular prey item or whether to move on in search of something better. The decision is based largely on time: how long it will take to find another prey item and, when another is found, how long it will take to capture that

Population Ecology: An Introduction to Computer Simulations. By Ruth Bernstein.

prey item in comparison with the one it has already found. These two components of foraging time are called search time and handling time.

Search time is inversely related to prey abundance: the more prey items, the less time it takes to find one. A shrew controls the abundance of its prey by controlling the number of prey species that it takes. If it takes only worms, its prey are less abundant and its search time longer than if it takes both worms and beetles. Handling time depends on the behavior of the prey. It takes less time to capture a worm than a beetle, because a beetle is more aware of the shrew's presence and can move faster.

In this model, the average time it takes to find a prey item is expressed as the inverse of the prey abundance

$$\text{time to find a prey item} = \frac{1}{\text{abundance of prey}} \tag{14.1}$$

where abundance is expressed as occurrence per second. For example, a prey abundance of 100 worms per square meter may be expressed as 3 worms per second. The advantage of converting density to occurrence per second is that it expresses abundance in time, which is the basis of this particular model. Adding handling time to the total foraging time, the equation becomes

$$\text{total time to find and capture a prey item} = \frac{1}{\text{abundance}} + \text{handling time} \tag{14.2}$$

which in MATLAB is written: `tiw=1/aw+hw`, where `w` stands for worm, `tiw` stands for the time it takes per prey item when only worms are acceptable, `a` stands for abundance, and `h` stands for handling. Similarly for the beetle: `tib=1/ab+hb`, where `b` stands for beetle, and `tib` stands for the time it takes per prey item when only beetles are acceptable. When a shrew chooses to eat only worms or only beetles, one of these formulas can be used to estimate foraging time.

Consider now the foraging time of a shrew if it decides to take *both* worms and beetles (`tiwb`). Now the total abundance of prey is $aw + ab$, and so the average time it takes to find something to eat is

$$\text{time to find a prey item} = \frac{1}{aw + ab} \tag{14.3}$$

In this model, the time it takes to capture a prey item, called the handling time, depends on whether the prey is a worm or a beetle. Thus, we need to answer this question: when a shrew encounters an acceptable prey item (worm or beetle), what is the probability that it is a worm? (The probability that an acceptable prey item is a beetle is just one minus the probability that it is a worm.) We do this by knowing the abundance of worms in relation to the abundance of all acceptable prey items – both worms and beetles:

$$\text{probability of being a worm} = \frac{\text{abundance of worms}}{\text{abundance of worms} + \text{abundance of beetles}} \tag{14.4}$$

which, in the language of our program, becomes

$$\text{probability of being a worm} = \frac{aw}{aw + ab}$$

The handling time for the worm component of the diet is then the probability that an encountered prey item is a worm times the handling time for a worm:

$$\text{handling time for worms in diet} = \frac{aw}{aw + ab}(hw) \tag{14.5}$$

We then use the same procedure for calculating handling time for a beetle:

$$\text{handling time for beetles in diet} = \frac{ab}{aw + ab}(hb) \tag{14.6}$$

Putting all these components together, the average time it takes a shrew to find and capture a prey item when foraging for both worms and beetles is

$$\text{total time to find and handle a prey item} = \frac{1}{aw + ab} + \frac{aw}{aw + ab}(hw) + \frac{ab}{aw + ab}(hb) \tag{14.7}$$

When translated into the language of MATLAB, equation (14.7) becomes

```
tiwb=1/(aw+ab)+hw*aw/(aw+ab)+hb*ab/(aw+ab)
```

Where `tiwb` indicates that both worms and beetles are acceptable prey items.

For this model, we assign worms a range of abundances from 0.005 to 0.03 per second, and assign beetles a range of abundances from half that of the worms, equal that of worms, to twice that of worms. We then see whether a shrew gets more food per unit time by foraging only for worms or by foraging for both worms and beetles under these various conditions of prey abundance. We assign the worm a handling time of 1 second and the beetle a handling time of 60 seconds.

First, graph the foraging time of a shrew taking only worms, where worm abundance varies from 0.005 to 0.03 per second (plotted in intervals of 0.0005 seconds):

```
>>tiw='1/aw + hw';
>>tiwplot=vectorize(tiw);
>>hw=1;
>>aw=0.005:0.0005:0.03;
>>figure
>>hold on
>>plot(aw,eval(tiwplot),'k')
```

This gives you a curve (in black), showing the average time per prey item (in seconds) for a shrew foraging for worms under different conditions of abundance.

Now, on the same graph, show the average time for a shrew foraging for both worms and beetles where the beetles are half as abundant as the worms:

```
>>tiwb='1/(aw+ab)+hw*aw/(aw+ab)+hb*ab/(aw+ab)';
>>tiwbplot=vectorize(tiwb);
>>hb=60;
>>ab=0.5*aw;
>>plot(aw,eval(tiwbplot),'b')
```

This curve appears in blue on your graph. Now see what the curve looks like when the abundance of beetles is equal to the abundance of worms:

```
>>ab=aw;
>>plot(aw,eval(tiwbplot),'g')
```

and then how it looks when there are twice as many beetles as worms:

```
>>ab=2*aw;
>>plot(aw,eval(tiwbplot),'r')
```

In the graph, the Y axis is average time (in seconds) for a shrew to find and capture a prey item, and the X axis is the abundance of worms (in worms per second). Label the graph, identifying each curve, and move it to a Microsoft Word document. Give the graph a figure legend.

In the simulation above, it took a shrew 60 seconds to capture a beetle as compared with 1 second to capture a worm. These handling times will change with environmental temperature, for a shrew is an endotherm (warm-blooded) whereas both the beetle and the worm are ectotherms (cold-blooded). As the temperature drops in late autumn, the shrew's mobility will remain the same but that of the worm and beetle will drop. Actually, the worm is already quite immobile so its activity will not change much. The main effect will be the decreased mobility of the beetle as its body becomes cooler.

Compare the optimal strategies that you obtained above with the optimal strategies when the environmental temperatures are cooler. Do this by making major changes in the handling time of the beetle. Label the graph, move it to a Word document, and write a figure legend.

What is the effect of temperature on the foraging strategy of a shrew? What if the predator were a frog preying on flies and/or butterflies? What if the predator were a snake preying on mice and/or shrews?

EXERCISE 15

Optimal Foraging: Searching Predators that Maximize Energy

The population density of most terrestrial predators is limited by their food supply. This means that individual predators who hunt more efficiently obtain more than their share of the food supply and leave more offspring than others in the population. Over the generations, the hunting behavior of the population should become more and more efficient with regard to time spent hunting, or energy expended in hunting, or both. This area of study is called optimal foraging theory.

There are two general categories of predators: searching and sit-and-wait. A searching predator moves through its habitat looking for suitable animals to eat. A sit-and-wait predator positions itself at a vantage point and waits for suitable prey to wander near. In this exercise, you will explore the behavior of a searching predator whose optimal foraging strategy is to maximize the energy that it gains during each foraging bout.

A predator that is itself in no particular danger while foraging, or does not have other pressing duties (like parental care), is likely to have a foraging strategy that maximizes energy return rather than a strategy that minimizes foraging time. Such a predator should choose prey according to (1) the energy content of the animal, (2) the energy spent in searching for the animal, and (3) the energy spent in capturing and eating it.

Consider a coyote deciding whether to hunt for mice or for rabbits. The energetic content of each prey item can be measured in a laboratory. Here, assume that a mouse contains 150 kilocalories and a rabbit contains 2000 kilocalories. The energy spent in searching for one of these animals is the number of calories expended per second while

Population Ecology: An Introduction to Computer Simulations. By Ruth Bernstein.

searching multiplied by the number of seconds spent in the search. The search time is inversely related to the abundance of the prey – the more abundant the prey population, the less time it takes to find individual prey:

$$\text{search time} = \frac{1}{\text{abundance of prey}} \tag{15.1}$$

The energy spent in searching is

$$\text{search energy} = \text{kilocalories burned per second while searching} \times \frac{1}{\text{abundance}} \tag{15.2}$$

In this exercise, our measure of abundance will be occurrence per second (the more abundant the prey, the more often it is encountered).

The final component to consider in this foraging strategy is the energy expended in capturing and killing the prey, which we will call the handling energy. The number of calories expended in handling the prey is equal to the calories burned per second multiplied by the number of seconds it takes to capture and kill the prey:

$$\begin{aligned}\text{handling energy} = {} & \text{kilocalories burned per second in handling the prey} \\ & \times \text{handling time in seconds}\end{aligned} \tag{15.3}$$

Putting all three components (caloric content, search energy, and handling energy) together, we get the equation for energy intake per prey item:

$$\begin{aligned}\text{energy intake per prey item} = {} & [\text{energy content}] - \left[(\text{search energy}) \times \left(\frac{1}{\text{abundance}}\right)\right] \\ & - [(\text{handling energy}) \times (\text{handling time})]\end{aligned} \tag{15.4}$$

in which search energy, abundance, handling energy, and handling time are all expressed in terms of seconds.

Putting this equation into the language of MATLAB, the net energy gain from eating a mouse is `eim=em-es/am-eh*hm`, where `eim` is the net energy intake from a mouse, `em` is the energy content of a mouse, `es` is the search energy, `am` is abundance, `eh` is the handling energy, and `hm` is the time it takes to handle the mouse. Similarly, the net energy gain when eating only rabbits is `eir=er-es/ar-eh*hr`.

The energy return for the strategy of taking *both* mice and rabbits is more complicated. Consider the coyote hunting for either a mouse or a rabbit. It encounters one of these animals. The probability that the encountered prey is a mouse (rather than a rabbit) depends on the abundance of mice relative to rabbits. If, for example, there are 90 mice for every 10 rabbits, then the probability that the encountered prey is a mouse is 90/100, or 0.9. In the language of MATLAB, this probability that the encountered prey is a mouse becomes

$$\frac{am}{am + ar} \quad (15.5)$$

and the probability that the prey is a rabbit is

$$\frac{ar}{am + ar} \quad (15.6)$$

For a coyote foraging on both mice and rabbits, the *average* energy content per prey item is thus

(energy in mouse × probability of mouse) + (energy in rabbit × probability of rabbit)

or

$$\text{average energy content per prey time} = (em)\left(\frac{am}{am + ar}\right) + (er)\left(\frac{ar}{am + ar}\right) \quad (15.7)$$

The energetic cost of searching is the number of calories expended per second while searching (es) times the number of seconds spent searching (which is inversely related to abundance of the two prey items):

$$\text{search energy} = (es)\frac{1}{(am + ar)} \tag{15.8}$$

The energetic cost of handling is the number of calories expended per second while handling (eh) the prey multiplied by the number of seconds it takes to handle (capture and kill) a prey multiplied by the probability that the prey is of a particular type (mouse or rabbit):

$$\text{handling energy} = (eh)(hm)\left(\frac{am}{am + ar}\right) + (eh)(hr)\left(\frac{ar}{am + ar}\right)$$

or

$$\text{handling energy} = eh\left[(hm)\left(\frac{am}{am + ar}\right) + (hr)\left(\frac{ar}{am + ar}\right)\right] \tag{15.9}$$

Putting all these energetic benefits and costs together, we get the formula for net energy gain when foraging on both mice and rabbits:

```
eimr=em*am/(am+ar)+er*ar/(am+ar)-es/(am+ar)-eh*
[hm*am/(am+ar)+hr*ar/(am+ar)]
```
(15.10)

Lastly, we need to develop commands for the total time it takes to find, catch, and kill each type of prey item. The time it takes to find a mouse or a rabbit is inversely related to its abundance; the time it takes to catch and kill is directly related to its handling time. The total time for a coyote hunting only mice is thus

$$\text{total time} = \frac{1}{\text{abundance}} + \text{handling time} \tag{15.11}$$

In MATLAB, Equation (15.11) is written

$$tim = \frac{1}{am} + hm$$

The total time for a coyote hunting both mice and rabbits is

$$timr = \frac{1}{am + ar} + hm\left(\frac{am}{am + ar}\right) + hr\left(\frac{ar}{am + ar}\right) \tag{15.12}$$

Use MATLAB to analyze the best strategy for a coyote foraging only on mice and then foraging on both mice and rabbits (omitting the only-rabbits diet). Begin with the time commands:

```
>> tim='1/am+hm';
>> timr='1/(am+ar)+hm*am/(am+ar)+hr*ar/(am+ar)';
>> timplot=vectorize(tim);
>> timrplot=vectorize(timr);
```

Then enter the energy commands:

```
>> eim='em-es/am-eh*hm';
>>eimr=['em*am/(am+ar)+er*ar/(am+ar)-es/(am+ar)-eh*(hm*am/
   (am+ar)+hr*ar/(am+ar))'];
```

(Enter the entire `eimr` equation on a single line in MATLAB.)

Now that you have the average time per item and the average energy per item for each of the three strategies, all you have to do is divide average energy by average time to obtain the average energy per time for each of the strategies. For the strategy of taking only mice, the average energy per time is

```
>>etm=symop(eim,'/',tim)
```

To see the formula written simply, type

```
>> pretty(etm)
```

Next, the strategy of taking both mice and rabbits:

```
>> etmr=symop(eimr,'/',timr)
>> pretty(etmr)
```

Prepare these functions for graphing by typing

```
>> etmplot=vectorize(etm);
>> etmrplot=vectorize(etmr);
```

Now assign values to the parameters. The mouse is much easier to catch and kill than the rabbit, so I will guess that it takes 6 seconds to handle a mouse and 120 seconds to handle a rabbit.

```
>> hm=6;
>> hr=120;
```

The energy content of a mouse is 150 kilocalories; the energy content of a rabbit is 2000 kilocalories.

```
>> em=150;
>> er=2000;
```

It probably costs a coyote about 1 kilocalorie per second to search and 5 kilocalories per second to capture and kill.

```
>> es=1.0;
>> eh=5.0;
>> figure
>> hold on
```

Starting with the strategy of taking only mice, the abundance of mice (`am`) will be plotted from 0.01 to 0.40 per second in steps of 0.01. (An abundance of 0.01 means that 1 mouse is seen every 100 seconds.) A vector with these values is generated with

```
>> am=0.01:0.01:0.40;
```

The curve, describing the foraging strategy of taking only mice, is plotted in black with

```
>> plot(am,eval(etmplot),'k')
```

Now consider the strategy of taking both mice and rabbits. Plot three curves, with the abundance of rabbits equal to 0.1 times the abundance of mice (blue), 0.5 times the abundance of mice (green), and the same as the abundance of mice (red).

```
>> ar=0.1*am;
>> plot(am,eval(etmrplot),'b')
>> ar=0.5*am;
>> plot(am,eval(etmrplot),'g')
>> ar=am;
>> plot(am,eval(etmrplot),'r')
```

The *X* axis is abundance, defined as occurrence per second. The *Y* axis is net energy gain. What is the optimal foraging strategy for the coyote? Label the graph, move it to a Microsoft Word document, and give it a figure legend.

The optimal strategy for the coyote may change, due to changes in the values of the parameters. In winter, for example, mice are easier to catch and kill because they are beneath the snow and do not see the coyote approaching. Repeat the simulation with a lower handling time for the mice (e.g., `hm=1`). What is the optimal strategy now? Label the graph, move it to a Word document, and give it a figure legend.

Design yet a third simulation, in which you change the value of a parameter and develop a scenario explaining the environmental condition that would produce this new value. Label the graph, move it to a Word document, and give it a figure legend.

EXERCISE 16

Optimal Foraging: Sit-and-wait Predators that Maximize Energy

The population density of most terrestrial predators is limited by their food supply. This means that individual predators who hunt more efficiently obtain more than their share of the food supply and leave more offspring than others in the population. Over the generations, the hunting behavior of the population should become more and more efficient with regard to time spent hunting, or energy expended in hunting, or both. This area of study is called optimal foraging theory.

There are two general categories of predators: searching and sit-and-wait. A searching predator moves through its habitat looking for suitable animals to eat. A sit-and-wait predator positions itself at a vantage point and waits for suitable prey to wander near. In this exercise, you will explore the behavior of a sit-and-wait predator whose optimal foraging strategy is to maximize the energy that it gains during each foraging bout.

The sit-and-wait predator remains in one place and makes decisions about when to pursue and when not to pursue the prey that it sees. A large part of this decision is how far the prey is from the predator. Consider a kingfisher sitting on its perch on a tree branch, gazing down at a river and deciding which fish to go for. For simplicity, assume that this foraging area forms a semicircle around the kingfisher and that all the fish are of the same size and behavior. When the kingfisher decides to take a particular fish, it dives from its perch, grabs the fish, and then returns to its perch.

In this exercise, you will determine the optimal size of the kingfisher's foraging area. Finding the outer boundary of the foraging area, beyond which the kingfisher should not

Population Ecology: An Introduction to Computer Simulations. By Ruth Bernstein.

pursue a fish, involves a trade-off between waiting and pursuit. Increasing the size of the foraging area decreases the time and energy spent waiting for a fish to appear (more fish to choose from) but increases the average time and energy spent in pursuit of the prey (longer distances to fly).

To find the optimal size of the foraging area, you need to know how the abundance of fish increases as the size of the semicircle increases. The area of a circle is πr^2, and so the area of a semicircle is $\frac{1}{2}\pi r^2$. The rate at which area changes with radius is the derivative of the area with respect to the radius, which is just πr. The total abundance of prey within a semicircle is obtained by integrating the abundance per unit area (e.g., per square meter) over the total area of the semicircle:

$$\text{abundance of prey in semicircle} = \int_0^{r_c} a(\pi r)dr \qquad (16.1)$$

where a stands for abundance per unit area and r_c stands for the cut-off radius (outer boundary of the semicircle).

The time that a kingfisher has to wait before a fish appears is then proportional to the reciprocal of this total abundance – the larger the area, the more fish, and the shorter the time that the kingfisher has to wait before it sees one. Defining abundance as number of fish that appear each second, we can define waiting time as

$$\text{waiting time} = \frac{1}{\int_0^{r_c} a(\pi r)dr} \qquad (16.2)$$

Let tw stand for waiting time (in seconds), a for abundance of the fish (as occurrence per second), r for radius of the semicircle, and rc for cut-off radius. Tell MATLAB to evaluate this equation by typing

```
>> tw=symop('1','/',int('a*pi*r','r','0','rc'))
>> pretty(tw)
```

(As you can see, the command pretty returns an equation that is easier to interpret.)

Graph the relation between waiting time and size of the foraging area (radius in meters). First, convert the equation for tw to a version that can be plotted on a graph:

```
>> twplot=vectorize(tw);
>> figure
>> hold on
```

Then assign values to the variables. I chose an abundance of 1 fish every 100 seconds (a=0.01) and foraging areas between 1 and 20 meters in radius, in steps of 0.1 meters.

```
>> a=0.01;
>> rc=1:0.1:20;
>> plot(rc,eval(twplot),'b');
```

On this graph, the X axis is radius of the foraging area (in meters) and the Y axis is waiting time. Keep the graph in MATLAB so that you can add to it later.

The time that it takes a kingfisher to travel from its perch to a fish is equal to the distance (as radius, in meters) divided by the velocity of flight (in meters per second): r/v. Thus, the time it takes a kingfisher to fly to a fish and then return to its perch is $2(r/v)$. The *average* pursuit time incorporates the time it takes to pursue fish in all parts of its foraging area, from the area right next to its perch ($r = 0$) to the outermost boundary ($r = rc$):

$$\text{average pursuit time} = \frac{\int_0^{r_c} \left(2\frac{r}{v}\right) a\pi r dr}{\int_0^{r_c} a\pi r dr} \qquad (16.3)$$

In this formula, $a\pi r dr / \int a\pi r dr$ is the number of fish at distance r divided by the total number of fish in the semicircle, which is the same as the probability that a fish appears at a distance r from the kingfisher. Integrating this probability times the pursuit time over all r from 0 to r_c gives the average pursuit time. This formula is entered in MATLAB as

```
>> tp=symop(int('a*pi*r*(2*r/v)','r','0','rc'),'/',int('a*pi*r','r','0','rc'))
>> pretty(tp)
```

Graph the relation between average pursuit time and size of the foraging area. Assign values of abundance and velocity. (I used an abundance of 1 fish per 100 seconds and a velocity of 1 meter per second.)

```
>> tpplot=vectorize(tp);
>> v=1;
>> plot(rc,eval(tpplot),'r')
```

The X axis of this graph is the radius (in meters) of the foraging area and the Y axis is time (in seconds). Label the graph, move it to a Microsoft Word document, and give it a figure legend. Why do these two curves differ in shape? What do they predict about foraging time for the kingfisher?

Now that we have formulas for the waiting time and average pursuit time, we can define the average time that it takes a kingfisher to get a fish as the sum of the waiting time and the average pursuit time:

```
>> ti=symop(tw,'+',tp)
>> pretty(ti)
```

The average energy gain that a kingfisher gets for every fish it eats is

$$\text{energy gain} = \text{energy intake} - \text{energy spent waiting} - \text{energy spent in pursuit} \quad (16.4)$$

where energy spent waiting is energy per second times the number of seconds, and energy spent in pursuit is energy per second times the number of seconds. Enter the equation in the language of MATLAB:

```
>> ei=symop('e','-','ew','*',tw,'-','ep','*',tp)
>> pretty(ei)
```

where `e` is the energy content of each fish, `ew` is the energy expenditure per second while waiting for a fish, and `ep` is the energy expenditure per second during pursuit.

Finally, the average rate of energy gain (kilocalories per second) from foraging in a semicircle of radius r_c is simply the average energy per fish divided by the average time spent getting the fish:

```
>> et=symop(ei,'/',ti)
>> pretty(et)
```

Graph the relation between energy gain per second and size of the foraging area. Assign values to the parameters. (I chose 100 kilocalories as the energy content of a fish, 0.1 kilocalories as the energy burned by a kingfisher while waiting, 2 kilocalories as the energy burned while pursuing, a flying speed of 1 meter per second, an abundance of 1 fish every 100 seconds, and a range of foraging areas with radius between 1 and 8 meters, in steps of 0.1 meter).

```
>> etplot=vectorize(et);
>> e=100; ew=0.1; ep=2; v=1;
>> rc=1: 0.1:8;
```

```
>> a=0.01;
>> figure
>> hold on
>> plot(rc,eval(etplot),'b')
```

On this graph, the X axis is the size of the foraging area (radius in meters) and the Y axis is the net energy gain (in kilocalories per second). To this graph, add curves predicting the optimal strategies for kingfishers hunting in other streams, where the fish populations are more dense and less dense (i.e., different *a* values). Label the graph, move it to a Word document, and write a figure legend. What is the effect of fish density on the optimal foraging strategy of a kingfisher?

What is the effect of environmental temperature on optimal foraging strategy of a kingfisher? See what happens when the temperature becomes colder and the energy costs of both waiting and pursuing times increase in this warm-blooded animal. Label the graph, move it to a Word document, and write a figure legend. Do you expect the optimal foraging area to differ in the winter from in the summer? In what way?

Next, see the effect of fish size on optimal size of the foraging area. Construct a new graph in which you place curves for various sizes of fish (kilocalories per fish, abbreviated *e*). Label the graph, move it to a Word document, and write a figure legend. What is the effect of size of fish on size of foraging area?

Lastly, see the effect of bad weather on optimal size of the foraging area, assuming that a kingfisher's flight velocity slows down with rain, sleet, and swirling winds. Try several flight velocities. Label the graph, move it to a Word document, and write a figure legend. How do you expect bad weather to modify the optimal size of a kingfisher's foraging area?

EXERCISE 17

Optimal Foraging: Pollinators

Bumblebee colonies consist of a queen and workers with different tasks. A forager is a worker with the single task of collecting nectar and pollen, and the quantity of this food determines the reproductive success of the colony. One expects, therefore, natural selection to have molded foragers to move from flower to flower in a very efficient manner. The study of how an animal should forage in order to maximize food intake is called optimal foraging theory.

Consider a bumblebee foraging on a lupine, a plant with long spikes of blue pea-like flowers. The lowest flower on the spike matures first and contains the most nectar. As the season progresses, this flower disintegrates and the next flower up becomes the lowest flower. And so the process continues throughout the flowering season. On any one day during the season, there is a linear relationship between floral position and caloric content of the flower. For a lupine, this relationship can be described as

$$\text{calories} = 20 - 1.7(i) \tag{17.1}$$

where i refers to the position of the flower (the bottom-most flower is number 1), 20 is number of calories per flower in the lowermost flower (the Y intercept), and 1.7 is the slope. Graph this relationship by entering

```
>> calories='20-1.7*i';
>> i=0:1:10;
```

Population Ecology: An Introduction to Computer Simulations. By Ruth Bernstein.

```
>> figure
>> plot(i,eval(calories))
```

On this graph, the X axis is the sequence of flowers on a lupine stalk, with 1 the lowest and 10 the highest, and the Y axis is the number of calories in the flower. Label the graph, move it to a Microsoft Word document, and write a figure legend.

A bumblebee tends to feed first from the bottom flower, which contains the most calories, and then work its way up the stalk. It leaves the plant, however, before it has fed on all the flowers. The question that you will answer, in this exercise, is: at what point along the stalk should the bumblebee leave, to seek the lowest flower of another stalk, if it is foraging in an optimal manner?

Suppose you have studied the energetics of this species of bee on a particular day in a particular meadow and have the following information:

P=probability that the plant has not been drained of nectar already=0.40

eb=energy expended in flying from one lupine plant to another=0.1 calorie

tb=time spent in flying from one lupine plant to another=5 seconds

ew=energy spent in flying from one flower to another on same stalk=0.03 calories

tw=time spent in flying from one flower to another on same stalk=0.05 seconds

ef=energy spent in emptying a "full" flower=0.02 calories

tf=time spent in emptying a "full" flower=15 seconds

ee=energy spent in sampling an "empty" flower=0.01 calories

te=time spent in sampling an "empty" flower=9 seconds

n=last flower position visited by a bee=at most 10

Write the following program and save it as *bumb* on your floppy disk:

```
function[netE]=bumb(P,eb,ew,ef,ee,tb,tw,tf,te,n)
netE=[];
sumi=0;
for i=1:n
  sumi=sumi+(-1.7*i+20);
  net=(P*sumi-eb-P*ew*(i-1)-P*i*ef-ee*(1-P))/...
    (tb+P*tw*(i-1)+P*i*tf+te*(1-P));
  netE=[netE net];
end;
```

The numerator in the line beginning with `net=` is the total energy gained by visits to i flowers. The denominator is the total time spent foraging on that plant and moving to the next plant. When the ratio is maximized, then the bee is foraging optimally. To run the program, enter values for the parameters:

```
>> cd a:
>> P=0.4; eb=0.1; ew=0.03; ef=0.02; ee=0.01;
>> tb=5; tw=0.05; tf=15; te=9; n=10;
>> [netE]=bumb(P,eb,ew,ef,ee,tb,tw,tf,te,n);
>> figure
>> hold on
>> plot(1:10,netE,'r')
```

On this graph, the X axis is the last flower (positioned between 1 and 10) visited before the bee leaves the stalk. The Y axis is the rate of net energy intake (calories per second). Label the graph, move it to a Word document, and write a figure legend. What is the optimal foraging strategy for a bumblebee under these conditions?

Explore the consequences of changing some of the parameters. Prepare a new graph for each variable that you change, placing on this graph the original curve (in red) as a sort of control. Distance between plants, for example, would vary from meadow to meadow. See how that affects the optimal strategy. Find the optimal strategy for a meadow that has a higher density of bees (i.e., P is higher). Find the optimal strategy for a different species of pollinator – a hummingbird, for example. Find the optimal strategy for a different species of plant – a penstemon, for example, which has a similar arrangement of flowers but a steeper decline in caloric content with height (e.g., change the slope from 1.7 to 2.5).

EXERCISE 18

Microparasite – Host Dynamics

The study of the spread of infectious diseases is an area of medicine known as epidemiology. Mathematical models of the interaction between host and pathogen are useful in predicting the rate at which a disease will spread and the expected level of disease in the population. This field of study, called mathematical epidemiology, is a subject of growing importance in medicine. The classic model for microparasites was developed, in 1927, by the British epidemiologists W. O. Kermack and A. G. McKendrick. A microparasite is an organism that can complete its life cycle within a single host. Most microparasites are viruses, bacteria, or fungi; a few are protists.

The Kermack–McKendrick model divides the host population into three compartments: susceptible, infected, and resistant (immune). For purposes of writing equations, we call these three subgroups x_1, x_2, and x_3 (see Figure 18.1). Individuals move from one group to another as the disease progresses.

The rate at which individuals move from the susceptible to the infected compartment is described by bx_1x_2, where b is the probability that contact between a susceptible and an infected individual produces another infected individual (related to invasion site and how the microbe travels) and x_1x_2 reflects the rate at which the two types of people come into contact (related to population density and social factors). The rate at which individuals move from infected to resistant (u) is inversely related to the length of the infectious period. For some diseases, resistant individuals can become susceptible again (e.g., the flu and the common cold), and so there is a parameter, called p, that describes this rate of flow.

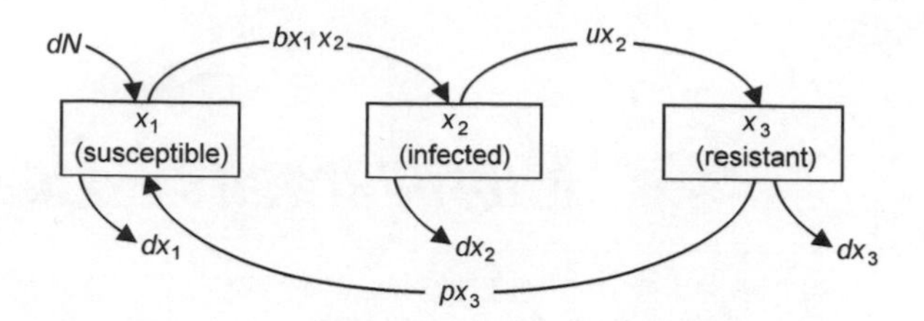

Figure 18.1 Flow diagram for the Kermack–McKendrick model of microparasite–host dynamics. Symbols: d = birth rate = death rate; b = probability that contact between a susceptible and an infected produces another infected; u = rate at which infected individuals become resistant; p = rate at which resistant individuals become susceptible

The Kermack–McKendrick model assumes zero population growth of the host population: the birth rate is equal to the death rate. Both rates are simply identified as d. The total number of individuals in the population (N) is the sum of the three x values. Thus, the number of newborns (all susceptible) is $d(N)$ and the number of deaths is $d(x_1 + x_2 + x_3)$. These parameters are shown in the flow diagram in Figure 18.1.

The population dynamics of each compartment are described by a differential equation. From the diagram, we can see that the change in number of individuals in the susceptible compartment is described mathematically as

$$\frac{dx_1}{dt} = -bx_1x_2 + px_3 + dN - dx_1 \tag{18.1}$$

The change in number of individuals in the infected compartment is described by

$$\frac{dx_2}{dt} = bx_1x_2 - ux_2 - dx_2 \tag{18.2}$$

and the change in number of individuals in the resistant compartment is described by

$$\frac{dx_3}{dt} = ux_2 - px_3 - dx_3 \tag{18.3}$$

Because the total population size is constant, we can substitute into the first equation (for x_1) a predictive equation for the resistant (x_3):

$$x_3 = N - x_1 - x_2 \tag{18.4}$$

$$\frac{dx_1}{dt} = -bx_1x_2 + p(N - x_1 - x_2) + dN - dx_1 \tag{18.5}$$

The parameters of the Kermack–McKendrick model vary from one disease to another, as shown in Table 18.1. In preparing this table, I assumed that the birth rate and the death rate equal 0.016 (the world death rate) except for infectious mononucleosis where I defined birth as entering high school and death as leaving high school, giving a birth rate and death rate of 0.25.

In this exercise, you will determine (for two diseases) (1) whether the pathogen will cause an epidemic, (2) whether the disease will persist (i.e., become endemic), (3) if the disease becomes endemic, what fraction of the population will be infected, and (4) when the epidemic will reoccur.

Begin this exercise by writing a program (in the program window) for the two differential equations:

```
function xdot=sir(t,x)
global n b p d u;
xdot(1,1)=-b*x(1)*x(2)+p*(n-x(1)-x(2))+d*n d*x(1);
xdot(2,1)=b*x(1)*x(2)-u*x(2)-d*x(2);
```

Save as *sir* on your floppy disk. (The Kermack–McKendrick model is also known as the SIR model, which stands for the three compartments: susceptible, infected, and resistant.)

Inform the computer you will be working from a program on your floppy disk and that n, b, p, d, and u are variables:

```
>> cd a:
>> global n b p d u;
```

Table 18.1 The Kermack–McKendrick parameters for specific diseases

Disease	b (transmission)	p	d	u (infectious period)
measles	0.01 (inhalation of droplets)	0.01	0.016	0.75 (14 days)
influenza	0.01 (inhalation of droplets)	1.00	0.016	0.95 (5 days)
whooping cough	0.01 (inhalation of droplets)	0.01	0.016	0.55 (28 days)
cholera	0.05 (swallowing feces)	0.10	0.016	0.80 (10 days)
leprosy	0.001 (touch)	0.01	0.016	0.01 (many years)
infectious mononucleosis	0.05 (swallowing saliva)	0.01	0.250	0.20 (50 days)

In the first part of this exercise, you will compare two of the diseases shown in the table. Choose a population size (I chose 1000) and then type in values of the parameters for the first disease.

```
>> n=1000; b=??; p=??; d=.016; u=??;
```

Construct a figure that traces the number of individuals in each of the three compartments over a period of 12 years. Begin with just one infected individual (no susceptibles $= n - 1$; no infected $= 1$). Apply the ordinary differential equation solver # 45 to your stored program *sir.*

```
>> figure
>> tspan=[0 12];
>> no=[n-1;1];
>> [t,x]=ode45('sir',tspan,no);
>> plot(t,x)
```

On this graph, the *X* axis is time and the *Y* axis is number of people (susceptible or infected). The number of susceptibles starts at 999. The number of infected starts at 1. (The number of resistants is not plotted; it is equal to the initial number (1000) minus the susceptibles minus the infected.) Label the graph, move it to a Microsoft Word document, and give it a figure legend. Repeat the simulation for another disease described in Table 18.1.

Now choose one of the diseases for further analysis. Suppose you are a physician and have to decide between a vaccine or an effective medicine for treating the disease. A vaccine converts susceptibles to resistants, passing very quickly through the infected stage. Thus, *b* and *u* are both very high and equal. For influenza, you need to make *b* and *u* very high and equal *and* reduce *p* (to around 0.01 – not all vaccines are effective). An effective medicine increases only *u*. Discuss the advantages and disadvantages of these two treatments.

Lastly, test the hypothesis that most infectious diseases of modern societies did not exist when we were hunter–gatherers (our natural ecology, for which we have evolved biological adaptations) because we lived in small societies. Choose one of the diseases, keeping all the parameters the same except the population size, to test this hypothesis. Run at least one simulation for a longer time, such as 100 years.

EXERCISE 19

Macroparasite – Host Dynamics

A macroparasite spends only part of its life cycle within one host individual. The rest of the cycle is spent either as a free-living individual or within an individual of another host species. Most macroparasites are arthropods (fleas, mites, lice, etc.) or worms (flatworms or roundworms). In this exercise, we will consider infections by worms.

Worms can infect any part of the body. The most common site is the small intestine, where they feed on digested materials. Worms rarely kill directly; instead, they weaken the host, reducing its reproductive output and increasing its vulnerability to other sources of mortality. The detrimental effect of worm infections increases with the worm burden (the number of worms within the host individual).

Ecologists and epidemiologists try to predict the long-term pattern of the interaction between worms and their hosts. Do the interactions contribute to the balance of nature? Can worms regulate the population density of their host? Can hosts regulate the population density of their parasite? In this exercise, you will explore the classic model of worm–host dynamics, which was developed in 1978 by Roy Anderson (a British epidemiologist) and Robert May (an Australian theoretical ecologist).

Anderson – May Model of Macroparasite – Host Dynamics

The Anderson–May model has three populations: the host population, the worm population within the hosts, and the worm population outside the hosts. The movement

Population Ecology: An Introduction to Computer Simulations. By Ruth Bernstein.

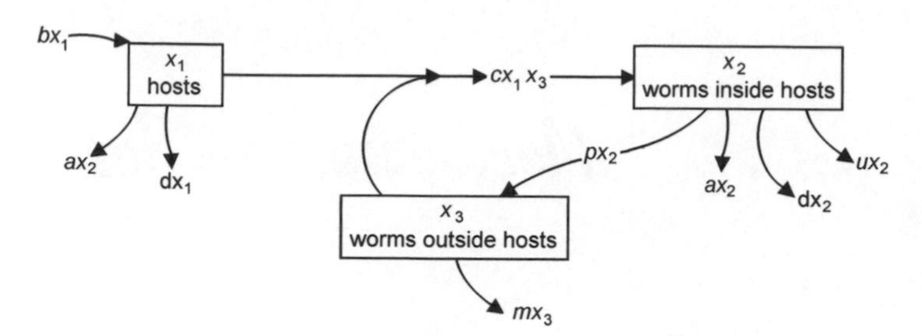

Figure 19.1 Flow diagram for the Anderson–May model of macroparasite–host dynamics. Symbols: $b =$ birth rate of host in absence of worms; $d =$ death rate of host in absence of worms; $a =$ effect of worms on host birth rate (decreases with larger worm burden) and host death rate (increases with larger worm burden); $c =$ rate at which worms enter a host from the environment; $u =$ natural mortality rate of adult worms in the hosts; $p =$ rate at which immature worms (usually eggs) are released into the environment; $m =$ death rate of immature worms in the environment outside the host

of worms from one population to another is shown in Figure 19.1. The parameters in this flow diagram control the population dynamics within the three populations, which are written as three differential equations:

$$\frac{\mathrm{d}x_1}{\mathrm{d}t} = (b - d)x_1 - ax_2 \tag{19.1}$$

$$\frac{\mathrm{d}x_2}{\mathrm{d}t} = cx_1x_3 - (u + a + d)x_2 - a[(k + 1)/k]x_2^{2/x_1} \tag{19.2}$$

$$\frac{\mathrm{d}x_3}{\mathrm{d}t} = px_2 - (m + cx_1)x_3 \tag{19.3}$$

The term $-a[(k + 1)/k]x_2^{2/x_1}$ in the equation describing population 2 controls the spatial distribution of the worms within the host population.

Write a program in MATLAB for simulating population growth in the three compartments, using the three differential equations. Open the program window and type

```
function xdot=worm(t,x)
global b d a c u p m k;
xdot(1,1)=b*x(1)-d*x(1)-a*x(2);
xdot(2,1)=c*x(3)*x(1)-(u+a+d)*x(2)-a*[(k+1)/k]*x(2)^2*1/x(1);
xdot(3,1)=p*x(2)-(m+c*x(1))*x(3);
```

Save the program as *worm* on your floppy disk.

Random Distribution of Worms within the Host Population

Surprisingly, the spatial distribution of worms within the host population has a large effect on the population size and stability. In this section, we consider a random distribution of worms. Various probability distributions can be used to describe, mathematically, the spatial distribution of a population. Anderson and May chose the negative binomial distribution, into which they incorporated three parameters: a (the negative effect of the worms on the host birth and death rates), x_3 (the number of worms in the environment at the beginning of the season), and k (the clumping factor):

$$\text{negative binomial} = \left(1 + a\frac{x_3}{k}\right)^{-k} \tag{19.4}$$

This term is added (along with a measure of variability, $2/x_1$) to the equation describing the dynamics of population 2. The parameter k determines the spatial distribution. In the equation, k is expressed as term $(k+1)/k$, which means that the larger the k, the closer the term is to 1, which would be a random distribution of worms in the host population. See how the negative binomial distribution approaches a random distribution (as described here by the Poisson distribution) for large values of k. The equations for the binomial distribution and the Poisson distribution predict the probability that a host *will not* be infected by a worm. Because it is easier to interpret the probability that a host *will* be infected, we simply subtract each equation from 1.

```
>> global r a c k
>> random='1-exp(-a*x3)';
>> a=0.006;
>> x3=0:10:1000;
>> figure
>> hold on
>> plot(x3,eval(vectorize(random)),'r');
>> negbinomial='1-(1+a*x3/k)^(-k)';
>> k=0.3;
>> plot(x3,eval(vectorize(negbinomial)),'k');
```

Repeat the negative binomial simulation with larger values of k (up to 30), using different colors for each curve. On this graph, the X axis is the number of worms in the environment at the beginning of the season and the Y axis is the probability that a host will become infected during the season. Label the graph, move it to a Microsoft Word document, and write a figure legend.

Using a random distribution of worms within the host population ($k = 30$), see how interactions among the three compartments influence the population sizes and stability. Choose a long time span (I chose 500 years) in order to detect long-term regulation if it exists. Choose any initial densities (I chose 100 for the host, 200 for worms inside the host, and 200 for worms outside the host). Because you are interested in whether or not worms can control host populations, make b greater than d so that the host population would grow if it were not for the effect of the worms. For the other parameters, I have chosen values that seem realistic.

```
>> cd a:
>> b=0.03; d=0.02; a=0.006; c=0.002; u=0.01; p=0.08; m=0.01; k=30;
>> tspan=[0 500];
>> no=[100;200;200];
>> [t,x]=ode45('worm',tspan,no);
>> figure
>> plot(t,x)
```

The X axis is the time in years. The Y axis is the size of each population (blue = hosts; green = worms inside hosts; red = worms outside hosts). Label the graph, move it to a Word document, and write a figure legend. Describe in words the dynamics of the three populations.

Now see if you can make them stable. Do this by changing the parameters (other than k) in ways that seem biologically realistic, using the flow diagram in Figure 19.1 as a guide. Do you find evidence that the worms control the host population? That the hosts control the worm population? After each simulation, label the graph, move it to a Word document, and write a figure legend. Describe the changes that you made in the model, their biological basis, and their effect on population stability.

Clumped Distribution of Worms within the Host Population

In a clumped distribution, some host individuals have many worms and others have few or no worms. The degree of clumping is described by the parameter k: the lower the k, the greater the degree of clumping. Using the same values for the parameters applied above, explore the effect of k on the population densities and stability. Begin with $k = 6$ (very little clumping).

```
>> k=6;
>> tspan=[0 500];
>> no=[100;200;200];
>> [t,x]=ode45('worm',tspan,no);
>> figure
>> plot(t,x)
```

Is the population stable? Label the graph, move it to a Word document, and give it a figure legend.

Repeat the commands, using smaller values of k (the worms are more clumped within the host population), going as low as $k = 0.3$. What do you conclude about the relation between k and population stability? Under what conditions did you achieve the greatest stability? Label the graphs, move them to a Word document, and give them figure legends.

It turns out that, in the real world, worms *are* clumped within their host populations. What do you think might produce this clumped dispersion pattern?

EXERCISE 20

Parasitoid–Host Dynamics

A parasitoid is an insect that lays its eggs in or on another insect or a spider, which is then consumed by the parasitoid larvae. Most parasitoids can parasitize only one species of host and only one particular stage of that host's development. The relationship is highly coevolved. The main challenges for the parasitoid are to find the host (usually by scent) and to synchronize its egg-laying with the host's life cycle. The main challenge for the host is to avoid being found.

More than 90 percent of all successful cases of biological control involve parasitoids controlling their host populations. Parasitoids are useful in controlling insect pests because they are highly specialized, which means that when their host species becomes scarce, they do not switch to new host species. They are particularly useful when they maintain their host population at a low density, which is better than causing extinction because if the host goes extinct, then the parasitoid also goes extinct. The host may then recolonize the area from somewhere else.

Parasitoid–host dynamics are governed primarily by the searching behavior of the parasitoid and the spatial distribution of the host. In this exercise, you will explore two models that differ in their description of the dispersion pattern of the host population. The first model is the negative binomial model, developed in 1978 by the Australian theoretical ecologist Robert May; it uses the negative binomial probability distribution to define the dispersion pattern. (The negative binomial distribution is also used in the model of macroparasite–host dynamics.) The second model is the Nicholson–Bailey model, developed in 1935 by A. J. Nicholson (an Australian entomologist) and V. A.

Population Ecology: An Introduction to Computer Simulations. By Ruth Bernstein.

Bailey (an Australian physicist); it uses the Poisson distribution to define the dispersion pattern.

The Negative Binomial Model of Parasitoid–Host Population Dynamics

This model is written in discrete time because the life cycles of both parasitoid and host are typically synchronized by the seasons. The population density of the host next year is the maximum rate of geometric growth (R_m) times the number of hosts in the population now (N_t) times the probability that an individual in the population will *not* be parasitized:

$$N_{t+1} = R_m N_t \left(1 + a\frac{P_t}{k}\right)^{-k} \tag{20.1}$$

The term in the parentheses is the probability of a host not being found. In addition to R_m, the parameters are: P, the number of parasitoids; a, the efficiency of the parasitoid in finding the host; and k, the clumping factor that describes the dispersion pattern of the host population. The population density of the parasitoid is predicted, from one year to the next, by the number of new parasitoids produced from each host (c) times the number of parasitoids now (P_t) times the probability that each parasitoid finds a host (one minus the probability that a host will not be found):

$$P_{t+1} = cN_t\left[1 - \left(1 + a\frac{P_t}{k}\right)^{-k}\right] \tag{20.2}$$

Begin the exercise by writing the following program in the program window:

```
function [n,p]=negbinom(runlen,no,po)
global r a c k
n=[no]; p=[po];
for t=1:runlen
    nprime=r*n(t)*(1+a*p(t)/k)^(-k);
```

```
    pprime=c*n(t)*(1-(1+a*p(t)/k)^(-k));
  if nprime <0 nprime=0; end;
  if pprime <0 pprime=0; end;
  n=[n nprime];
  p=[p pprime];
end
```

Save the program as *negbinom* on your floppy disk. In the command window, enter the two equations for finding equilibrium values:

```
>> cd a:
>> global r a c k
>> nprime='r*n*(1+a*p/k)^(-k)';
>> pprime='c*n*(1-(1+a*p/k)^(-k))';
```

Find the equations that describe zero population growth for the two interacting populations:

```
>> [nhat,phat]=solve(symop(nprime,'-','n'),symop(pprime,'-','p'),'n,p');
>> nhat=simple(nhat)
>> phat=simple(phat)
```

The equations in the second row and first column `(2,1)` are appropriate for our use.

See the effect of the parameters k, a, and c on the population densities. Begin with the effect of k. (Reasonable values are between about 3 and 0.3 – remember that the lower the k, the greater the degree of clumping.) Run each simulation for about 50 years and put all the simulations involving a change in k on the same graph. Do this by first typing

```
>> figure
>> hold on
```

Using the commands below, place the equilibrium point (an asterisk) on the graph and then start the populations at some densities other than the equilibrium values. I suggest that you write these commands in the program window in order to avoid having to retype them. Open the program window and type

```
r=2; a=0.001; c=1;
k = 2;
nhat_n=eval(sym(nhat,2,1));
phat_n=eval(sym(phat,2,1));
plot(nhat_n,phat_n,'*')
[n p]=negbinom(50,1.01*nhat_n,1.01*phat_n);
plot(n,p,'r')
```

Save these commands on your floppy disk as parasitoid. Run the simulation (by entering parasitoid) with different values of *k* and different colors or line styles. On this graph, the X axis is the number of hosts and the Y axis is the number of parasitoids. Label the graph, move it to a Microsoft Word document, and write a figure legend. What is the effect of *k* on stability? What is the effect of *k* on equilibrium densities? If the host is an insect pest, what can you advise the farmer to do to achieve the best dispersion pattern of the insect pest?

Next, type

```
>> figure
>> hold on
```

and then repeat the commands for a new set of simulations in which you keep *k* constant and vary *a*. Label the graph, move it to a Word document, and write a figure legend. If the host is an insect pest that you want to maintain in low densities, which *a* would you choose? What can you advise the farmer to do to achieve this particular *a* efficiency of the parasitoid in finding the host?

Now type

```
>> figure
>> hold on
```

and repeat the commands for a new set of simulations in which you vary only *c*. Label the graph, move it to a Word document, and write a figure legend. What aspects of the host and of the parasitoid influence the number of new parasitoids produced from each host?

The Poisson Model: Controlling an Insect Pest

Suppose you are assigned the task of helping corn farmers rid their fields of the corn borer, which is the larval stage of a European moth. This moth, which consumes mainly the stems of corn plants, has been introduced into the United States from Europe. You approach the problem by using a computer model to explore the effect of various kinds of parasitoids on the corn-borer population. We use here the Nicholson–Bailey model that is modified to incorporate logistic growth. This particular model of parasitoid–host dynamics assumes that the probability of a host *not* being found follows a Poisson distribution. The Poisson distribution resembles the negative binomial distribution with a relative large *k* (the clumping factor).

$$\text{Poisson: probability that a host will } not \text{ be detected} = e^{-aP} \tag{20.3}$$

where a is the searching efficiency of the parasitoid and P is the number of parasitoids. Recall that the negative binomial is

$$\text{negative binomial: probability that a host will } not \text{ be detected} = \left(1 + a\frac{P}{k}\right)^{-k}$$

Compare the Poisson distribution with the negative binomial distribution, using values of k between 0.5 and 8.

```
>> poisson='exp(-a*p)';
>> negbinomial='(1+a*p/k)^(-k)';
>> a=0.001;
>> p=0:100:10000;
>> figure
>> hold on
>> plot(p,eval(vectorize(poisson)),'r')
>> k=0.5;
>> plot(p,eval(vectorize(negbinomial)),'g')
```

Repeat the last two commands for larger k values. On this graph, the X axis is the number of parasitoids and the Y axis is the probability that a host will *not* be discovered by the parasitoid. Move this graph to a Word document and write a figure legend. How do the two probability distributions differ?

When the Poisson distribution is substituted for the negative binomial distribution, we get the following equations for the host and the parasitoid:

$$N_{t+1} = R_m N_t(e^{-aP_t}) \tag{20.4}$$

$$P_{t+1} = cN_t(1 - e^{-aP_t}) \tag{20.5}$$

Adding a carrying capacity to the host population, we get

$$N_{t+1} = RN_t\left(1 - \frac{N_t}{K}\right)(e^{-aP_t}) \tag{20.6}$$

Enter the following computer program for the Nicholson–Bailey model in which the host has a carrying capacity:

```
function [n,p]=nbk(runlen,intro,no,po)
global r a c k
n=[no]; p=[0];
for t=1:runlen
  if t<intro
    nprime=r*n(t)*(1-n(t)/k);
    if nprime<0
      nprime=0;
    end;
  n=[n nprime];
  p=[p 0];
  if t==(intro -1)
    avg1=mean(n);
    maxload1=max(n);
  end;
  else
  if t==intro
    p(t)=po;
  end;
  nprime=r*n(t)*(1-n(t)/k)*exp(-a*p(t));
  pprime=c*n(t)*(1-exp(-a*p(t)));
  if nprime <0
    nprime=0;
  end;
  if pprime<0;
    pprime=0;
  end;
  n=[n nprime];
  p=[p pprime];
  if t==runlen
    avg2=mean(n(intro:runlen));
    maxload2=max(n(intro:runlen));
    avgdiffer=(avg1-avg2);
    maxdiffer=(maxload1-maxload2);
    end;
  end;
end;
```

Save the program as *nbk* (i.e., Nicholson–Bailey with *K*) on your floppy disk.

Now, run your first biocontrol experiment. Start the simulation without the parasitoid and observe the pattern of growth of the host in the absence of its enemy. Then introduce the parasitoid and see what happens to the host population. In these equations, *r* is the geometric rate of population growth of the corn borer, *a* is the efficiency of the parasitoid in finding the corn borer, *c* is the number of offspring produced by the parasitoid from each corn borer that is parasitized, and *k* is the carrying capacity of the corn borer. To begin, I suggest you choose the same values that I did and then modify them in later simulations. Run the entire experiment for 100 years, introduce the parasitoid after 20 years, begin with 2000 corn borers and 800 parasitoids. (You may want to write the commands in your program window so you do not have to retype them for each simulation.)

```
>> cd a:
>> r=3; a=0.01; c=1; k=8000;
>> [n,p]=nbk(100,20,2000,800);
>> figure
>> hold on
>> plot(0:100,n,'g')
>> plot(0:100,p,'r')
```

On this graph, the X axis is time (in years) and the Y axis is population density. The green curve is the density of corn borers; the red curve is the density of the parasitoid.

Now, on different graphs, try several different species of parasitoid, with different capacities for finding the corn borer (*a*) and for using it to produce more parasitoids (*c*). Always change just one variable at a time so that you know the effect of each variable. Find the best parasitoid that you can within the time that you have. The best parasitoid is one that keeps its host at the lowest possible density (other than zero) yet does not go extinct itself.

EXERCISE 21

Conserving an Endangered Species

Understanding the dynamics of a population is an essential first step in preventing extinction of an endangered species. You now have the tools to explore, by means of computer simulations, the dynamics of an endangered species and to predict the most effective way of reversing the decline in numbers. Choose one of the following four endangered species to manage: the Violet-necked Duck, the Brown-spotted Toad, the Niwot Mouse, or Sally's Golden Butterfly. Your conservation project should consist of the following steps:

(1) Construction of a life table

(2) Estimation of the rate of decline

(3) Sensitivity analysis

(4) Development of a management plan

(5) Application of the management plan

(6) Prediction of recovery time

Construction of a Life Table

This part of your job requires a team of good field biologists and takes several years to complete. Assume here that the fieldwork has been done. The data for each of the four species are presented in the final section.

Estimation of the Rate of Decline

To convince yourself and your supporters that the population is declining, you need an initial estimate of the population growth rate so you can predict the population size at some time in the future. A statistic that is easy for the public to understand is how long it will take for the population to become half the size it is now. To obtain this statistic, first calculate the net replacement rate, R_0; the generation time, T; and the continuous rate of growth r from information in the life table. (While the growth of the population is discrete, the result is the same whether you estimate by using r or R.)

$$R_0 = \sum l_x m_x \tag{21.1}$$

$$T = \frac{\sum x l_x m_x}{\sum l_x m_x} \tag{21.2}$$

In the population growth equation $N_t = N_0 e^{rt}$, substitute generation time for generic time:

$$N_t = N_0 e^{rT} \tag{21.3}$$

$$\frac{N_T}{N_0} = e^{rT}$$

Apply the definition of R_0:

$$\frac{N_T}{N_0} = R_0 = e^{rT}$$

Manipulate to get r:

$$\ln R_0 = rT$$

$$\frac{\ln R_0}{T} = r \tag{21.4}$$

Substitute your values of R_0 and T to obtain r. Return to the basic population growth equation to calculate how long it will take the population to decrease to half its present size:

$$N_t = N_0 e^{rt}$$

$$\frac{N_t}{N_0} = \frac{1}{2} = e^{rt} \qquad (21.5)$$

$$\ln 0.5 = rt$$

$$\frac{\ln 0.5}{r} = t$$

The solution to this equation is the number of years (or weeks for a butterfly) it will take for the population to become half the size that it is now.

Sensitivity Analysis

A sensitivity analysis determines which part of the life history has the greatest effect on the population growth rate. Once this factor (age-related mortality or fecundity) is known, a management program can be aimed directly at that factor.

The first step in a sensitivity analysis is to develop a Leslie matrix, as described in Exercise 3. Do this by calculating p_x and F_x values from the life table and arranging them in an age-structured matrix.

The next step in a sensitivity analysis is to alter each aspect of the matrix and see the effect on the population growth rate R. As the number of offspring produced by each adult female does not vary much, we restrict our analysis to the survival values. To be sure that you detect subtle differences in the effects of mortality on population growth, you need to change, one by one, each p_x value to 1. It is unlikely that your management program will be able to achieve 100 percent survival from one age to the next, but this analysis will tell you what would happen to the population growth rate if you could. *Remember that if you change a p_x value prior to an age of reproduction, this change will alter the F_x value as well – be sure to change both*. Also keep in mind that *each time you change a value of p_x or F_x, you need to change it back to the original value* before you evaluate the effects of another p_x.

To enter the Leslie matrix into MATLAB, type in the values row by row, with a semicolon between rows:

```
>> m = [x x x x x x x;
        x x x x x x x;
```

```
xxxxxxx;
etc.]
```

Then ask the program to give you the largest eigenvalue, which is the geometric rate of growth (R) for a population undergoing discrete growth. This estimate of R is especially good because it is for a population with a stable age distribution.

```
>> R=max(max(abs(eig(m))))
```

Now, raise the first p_x value to 1. This is done by identifying its position in the matrix, row 2 and column 1:

```
>> m(2,1)= ;
```

If this change in p_x affects the F_x value (position 1,2) as well, then enter its new value:

```
>> m(1,2)= ;
>> R=max(max(abs(eig(m))))
```

Return the parameters to their original values. Find R again, just to check that you have restored your original matrix. This R should be the same as your first one.

Repeat the same procedure as you explore the sensitivity of each age, raising just one value of p_x (and associated F_x) at a time. Keep a record of the change that you make and of the new R value. Which change produces the highest R?

Development and Application of a Management Plan

Now that you know which age has the greatest effect on the population growth rate of the endangered species, as well as the major source of mortality at that age (see the life table for the species that you chose), you can develop a management plan for recovery. Use an appropriate model, chosen from the ones available in this text for predators, macroparasites (worms and the malaria protozoan), and microparasites (viruses). Apply the known parameters (given in the life table) and then modify these parameters in realistic ways to help the population recover. (Be sure to apply the correct rates of population growth: continuous or discrete, maximum or actual.) Once you have decided how to change the parameters, describe specifically how you will carry out your plan in the field.

Prediction of Recovery Time

Suppose that you completely eliminate the mortality factor having the greatest effect on the species under study. Use this new R (the highest one that you achieved) to describe, graphically, the population size over the next 20 years (or weeks for a butterfly). Use the equation for logistic growth, discrete version, described in Exercise 6.

What would be the effect of eliminating this mortality factor on other species in the food web?

The Four Endangered Species: Field Data

The Violet-necked Duck

The Sawhill ponds are home to 100 Violet-necked Ducks. The carrying capacity is approximately 1000. The maximum rate of population growth (the geometric rate for discrete growth) is estimated to be 1.5 (i.e., under ideal conditions, the population could increase by 50 percent each year). In addition, you have a life table (Table 21.1) and the following information about Violet-necked Ducks.

Between the time of birth and the first year of age, a duck goes from an embryo in an egg to a nestling and then to a fledgling. Because Violet-necked Ducks nest on the ground, the ducklings are vulnerable to predation by raccoons. Between ages 1 and 2, the ducklings begin to investigate their environment and then to mate for the first time. As they move through the habitat, they suffer from predation by red foxes. By 2 years of age, most ducks have accumulated significant numbers of intestinal worms to weaken their bodies; many die because they cannot survive the physical stress of migration. These worms pass from duck to duck through the water via a freshwater mollusk. By age 4, the immune system begins to decline and the ducks are especially vulnerable to dysentery, which spreads from duck to duck as they feed in water that is contaminated with their own feces. At 5 years of age, many ducks acquire malaria, a protozoan that is spread by way of mosquitoes. (Because this protozoan needs a secondary host – the mosquito – to complete its life cycle, it is a macroparasite.) Thereafter, their bodies are so deteriorated that they soon die of old age.

Table 21.1 Life table for the Violet-necked Duck

x	l_x	p_x*	m_x	l_xm_x	F_x**	*Major cause of mortality* (for descriptions of models and variables, see the relevant exercises in this text)
0	1		0			predation by raccoons ($a = 0.4$; $b = 0.01$; $d = 0.25$)
1	0.5		1			predation by red foxes ($a = 0.6$; $b = 0.02$; $d = 0.1$)
2	0.1		2			intestinal worms ($b = 0.03$; $d = 0.02$; $a = 0.01$; $c = 0.2$; $u = 0.001$; $p = 0.02$; $m = 0.01$)
3	0.03		3			intestinal worms ($b = 0.03$; $d = 0.02$; $a = 0.005$; $c = 0.3$; $u = 0.001$; $p = 0.03$; $m = 0.01$)
4	0.009		4			dysentery (a bacterium) ($b = 0.07$; $p = 0.01$; $u = 0.8$; $d = 0.06$)
5	0.0027		6			malaria (macroparasite) ($b = 0.05$; $d = 0.02$; $a = 0.03$; $c = 0.2$; $u = 0.01$; $p = 0.01$; $m = 0.01$)
6	0.00054	0	9			old age

*$p_x = l_{x+1}/l_x$
**$F_x = p_{x-1}m_x$

Table 21.2 Life table for the Brown-spotted Toad

x	l_x	p_x*	m_x	l_xm_x	F_x**	*Major cause of mortality* (for descriptions of models and variables, see the relevant exercises in this text)
0	1		0			predation by salamanders ($a = 0.2$; $b = 0.05$; $d = 0.15$)
1	0.4		0			predation by gray jays ($a = 0.8$; $b = 0.02$; $d = 0.2$)
2	0.08		0			chytrid fungus (microparasite: $b = 0.005$; $p = 0.01$; $u = 0.001$; $d = 0.05$)
3	0.016		40			chytrid fungus (same as above)
4	0.0048		40			chytrid fungus (same as above)
5	0.00144	0	40			old age

*$p_x = l_{x+1}/l_x$
**$F_x = p_{x-1}m_x$

The Brown-spotted Toad

The Thompson Wetlands are home to 50 Brown-spotted Toads. The carrying capacity is approximately 500. The maximum rate of population growth (the geometric growth rate for discrete growth) is estimated to be 1.5 (i.e., under ideal conditions, the population could increase by 50 percent each year). In addition, you have a life table (Table 21.2) and the following information about this population of toads.

Between ages 0 and 1, the eggs hatch and develop into tadpoles, which are eaten by salamanders. Near the end of the first year, the tadpoles metamorphose into toadlets and disperse away from their natal site. At this time, they are vulnerable to predation by gray jays, which are particularly abundant because their food supply in winter is supplemented by hand-outs from skiers and snowshoers. Between ages 2 and 3 the toads overwinter for the first time in underground burrows, where an introduced disease – the chytrid fungus – is passed from one toad to another. This fungus grows on the mouth parts and prevents feeding. It kills every toad that it infects.

The Niwot Mouse

The Niwot Mouse lives in riparian habitats in the short-grass prairie near the town of Niwot. You estimate that the population size is only about 60, but the carrying capacity is around 500. The maximum rate of population growth (the geometric growth rate for discrete growth) is estimated to be 1.5 (i.e., under ideal conditions, the population could increase by 50 percent each year). In addition, you have a life table (Table 21.3) and the following information about this population of mice.

Between the time of birth and the age of 1 year, the young remain near the nest where they are especially vulnerable to weasels, which are small enough to burrow into the leaf litter or ground where the nests are located. At the age of 1, the juveniles begin to disperse in search of their own nesting area. At this time, their most lethal enemy is a virus transmitted to the mice via the saliva of domestic cats (which often "mouth" the mice), and then from one mouse to another via their saliva. This virus, introduced two years ago, kills a mouse within 10 days. At the age of 2, the mice begin to reproduce and to acquire intestinal worms that cause them to weaken and die. Both species of worms move between the mouse and a snail by way of worm eggs or worm larvae on the vegetation. Both the mice and the snails eat the vegetation.

Sally's Golden Butterfly

Sally's Golden Butterfly lives in a large meadow between two rivers. You estimate that there are 200 left, and that the carrying capacity is around 8000. The maximum rate of population growth (the geometric rate for discrete growth) is estimated to be 1.20 per week (under ideal conditions, the population could increase 120% each week). In addition, you have a life table (Table 21.4), which gives a 7 by 7 Leslie matrix, and the following information about this butterfly population.

The female of this species of butterfly lays her eggs on tulip–gentian plants. The eggs hatch in late May and the tiny larvae feed, during their first week of life, on the leaf buds of the tulip–gentian. The major source of mortality during this time is bad weather – late spring snow and sleet. When they are 1 week old, the young caterpillars feed on new leaves of the tulip–gentian and are parasitized by torymid wasps; each wasp lays just one

Table 21.3 Life table for the Niwot Mouse

x	l_x	p_x*	m_x	$l_x m_x$	F_x**	*Major cause of mortality* (for descriptions of models and variables, see the relevant exercises in this text)
0	1		0			predation by weasels ($a = 0.2$; $b = 0.05$; $d = 0.15$)
1	0.5		1			cat distemper virus, spread by saliva ($b = 0.08$; $p = 0.001$; $u = 0.8$; $d = 0.08$)
2	0.1		2			intestinal worms ($b = 0.05$; $d = 0.01$; $a = 0.01$; $c = 0.005$; $u = 0.01$; $p = 0.08$; $m = 0.01$)
3	0.03		3			intestinal worms ($b = 0.03$; $d = 0.01$; $a = 0.05$; $c = 0.003$; $u = 0.01$; $p = 0.03$; $m = 0.01$)
4	0.006	0	3			old age

*$p_x = l_{x+1}/l_x$
**$F_x = p_{x-1}m_x$

Table 21.4 Life table for Sally's Golden Butterfly

x	l_x	p_x*	m_x	$l_x m_x$	F_x**	*Major cause of mortality* (for descriptions of models and variables, see the relevant exercises in this text)
0 (hatchling)	1		0			cold, wet weather
1 (1–2 mm)	0.7		0			parasitoid (torymid wasp) (negative binomial: $a = 0.001$; $c = 1$; $k = 0.3$)
2 (2–4 mm)	0.42		0			parasitoid (chalcid wasp) (negative binomial: $a = 0.002$; $c = 2$; $k = 1$)
3 (4–6 mm)	0.168		0			predatory bird (horned larks) ($a = 0.3$; $b = 0.003$; $d = 0.07$)
4 (6–8 mm)	0.084		0			parasitoid (round-headed fly) (negative binomial: $a = 0.0015$; $c = 2$; $k = 1.5$)
5 (pupa)	0.0252		0			parasitoid (chalcid wasp) (negative binomial: $a = 0.02$; $c = 4$; $k = 2$)
6 (adult)	0.005	0	80			genetically programmed death after reproduction

*$p_x = l_{x+1}/l_x$
**$F_x = p_{x-1}m_x$

egg in each caterpillar that it finds. When the caterpillars are 2 weeks old, they are parasitized by chalcid wasps, which lay two eggs in each caterpillar that they find. When the caterpillars are 3 weeks old, they begin to feed on more exposed parts of the tulip–gentian plants and suffer high mortality from horned larks, which feed caterpillars to their newly hatched chicks during mid-June. When the caterpillars are 4 weeks old, the round-headed fly lays its eggs (two per host) in them. At 5 weeks of age, they become pupae and yet another parasitoid – a chalcid wasp of a different species – lays its eggs (four per host) in each pupa that it finds. At 6 weeks, the adult emerges from the pupa, lays its eggs, and dies.

EXERCISE 22

Controlling an Invasive Species

Conservation biology involves not only conserving endangered species, but also controlling invasive species that have colonized an area where they are not coevolved components of the food web. Understanding the dynamics of a population is an essential first step in controlling an invasive species. You now have the tools to explore, by way of computer simulations, the dynamics of such a species and to predict the most effective way of reversing its rapid growth and expansion. Choose one of the following four invasive species to manage: the Brown Tree Snake, the Starling, the Gray Squirrel, or the Asian Longhorned Beetle.

Suppose the invasive species that you are planning to control has recently arrived in an area and is causing a decline in some of the native species. An environmental agency offers you $500 000 to slow its spread throughout the region, and an additional $500 000 if you completely halt the spread. You accept the challenge. Your intervention program should consist of the following steps:

(1) Construction of a life table

(2) Estimation of the present rate of spread

(3) Sensitivity analysis to determine which age group to focus on

(4) Intervention

(5) Prediction of rate of spread after intervention

(6) Further intervention, if necessary and you decide to earn the extra $500 000.

Population Ecology: An Introduction to Computer Simulations. By Ruth Bernstein.

Construction of a Life Table

You hire a team of field biologists to study the life history of this species. The field data for each of the four invasive species are presented in the final section.

Estimation of the Present Rate of Spread

Before the government pays you any money, it wants to know how fast the species is spreading now. Use the computer program *inwave* described in Exercise 2 to graph the spread of the population through time. Decide on a habitat length, a run length and the number needed to establish a population. Calculate the rate of population growth and the rate of dispersal, as described below.

The geometric rate of increase, R, is calculated from information in the life table. To obtain R, first calculate R_0, T and r:

$$R_0 = \sum l_x m_x \qquad (22.1)$$

$$T = \frac{\sum x l_x m_x}{\sum l_x m_x} \qquad (22.2)$$

In the population growth equation $N_t = N_0 e^{rt}$, substitute generation time for generic time:

$$N_T = N_0 e^{rT} \qquad (22.3)$$

$$\frac{N_T}{N_0} = e^{rT}$$

Apply the definition of R_0:

$$\frac{N_T}{N_0} = R_0 = e^{rT}$$

Manipulate to get r:

$$\ln R_0 = rT$$

$$\frac{\ln R_0}{T} = r \tag{22.4}$$

Then apply the equality

$$R = e^r \tag{22.5}$$

To find the dispersal rate, calculate how many individuals will be in the population next year (or week for a beetle):

$$N_{t+1} = RN_t \tag{22.6}$$

Before reproduction, the population is at its carrying capacity. After reproduction, all individuals in excess of the carrying capacity will disperse. Thus, for example, if the carrying capacity were 100 and R were 1.5, then the number of individuals next year (or week) would be 150, and the number that would disperse is 50. The *proportion* of the population that disperses (d) is the number of dispersers divided by the population total prior to dispersal (e.g., $50 \div 150$). Every time that R changes, d changes as well.

Use these values of R and d in the program *inwave*. Graph the present spread of the population as estimated from your value of R calculated from the life table information. Use the command `hold on` to allow later additions to the figure. Your field biologists have measured the average distance moved by a dispersing individual and found it to be x kilometers. (Choose, here, an x that seems reasonable for the particular species that you are studying.) Thus, the Y axis of your graph should read "Distance (times x kilometers)."

Sensitivity Analysis

A sensitivity analysis determines which part of the life history has the greatest effect on R, which, in turn, determines the rate of spread. Once this age group is identified, a management plan can be aimed directly at its mortality factors. The first step in the sensitivity analysis is to develop a Leslie matrix, as described in Exercise 3. Do this by calculating p_x and F_x values from the life table and arranging them in an age-structured matrix. To enter the Leslie matrix into MATLAB, type in the values row by row, with a semicolon between rows:

```
>> m = [x x x x x x x;
        x x x x x x x;
        x x x x x x x;
        etc. ]
```

Then ask the program to give you the largest eigenvalue, which is the geometric rate of growth (R) for a population undergoing discrete growth. This estimate of R is especially good because it is for a population with a stable age distribution.

```
>> R=max(max(abs(eig(m))))
```

The next step in a sensitivity analysis is to alter each p_x value and see its effect on the population growth rate R. It does not make sense to make a p_x value equal to 0, because then there would be no older age groups. Instead, halve each p_x value and see the effect on R. *Remember that each time you change a p_x value, it changes the F_x for the following age*. Also, keep in mind that each time you change a p_x value and record the new R, you need to restore the Leslie matrix to its original form before you try the next p_x value.

Begin by reducing the first p_x value by one-half. This is done by identifying its position in the matrix, row 2 and column 1:

```
>> m(2,1)= ;
```

If this change in p_x affects the F_x value (position `1,2`) as well, then enter its new value:

```
>> m(1,2)= ;
>> R=max(max(abs(eig(m))))
```

Return the parameters to their original values. Find R again, just to check that you have restored your original matrix. This R should be the same as your first one.

Repeat the same procedure as you explore the sensitivity of each age, lowering just one value of p_x (and associated F_x) at a time. Keep a record of the change that you make and of the new R value. Which change produces the lowest R?

Intervention

Now that you know which stage of life has the greatest effect on the spread of the invasive species, you can develop a plan for slowing the rate of spread. What will you do? Answer this question simply by describing in words how you will halve the p_x for the most sensitive age group. You can modify an existing enemy or you can use humans acting as enemies.

Prediction of Rate of Spread after Intervention

Add to your previous graph a curve showing the rate of spread predicted by using your most precise estimate of R (from the unmodified Leslie matrix, which is the R after a stable age distribution has been reached). Use different colors or line styles for all these curves. How does this R compare with your estimate from the life table? Now, on the same graph, show the predicted rate of spread after application of your intervention. Describe to the government's conservation agency the rate of spread (in kilometers) of the invasive species under all three conditions. Keep the graph for a later addition.

Further Intervention

Have you been able to slow the spread of the invasive species? Have you been able to halt the spread altogether? If not, what else can you do to prevent the pest from spreading throughout the region?

Do another sensitivity analysis, *keeping the most sensitive* p_x (determined in your first analysis) *at half its original value* while halving, one by one, the other values of p_x. Find out which age group now contributes most to the population growth rate. Using the lowest R value that you find, place a final curve on the graph of the spread of the species through the region. (If R is 1 or less, then d is equal to 0.)

If you reduce the population growth rate to an R of less than 1, then the invasive species will remain below carrying capacity and there will be no further spread of the population. Use one of the programs (predator–prey, microparasites, or macroparasites), as indicated by your most sensitive age group (from your first sensitivity analysis), to simulate how you will intervene further to maintain the population at densities of less than its carrying capacity. Use the R value that you obtained after your first intervention (the lowest R in the first sensitivity analysis) to see the effects of the second intervention.

The Four Invasive Species: Field Data

The Brown Tree Snake

The Brown Tree Snake (*Coluber constrictus*) is a form of "racer" that is a pest on Guam Island where it arrived in the 1940s as a few young snakes that somehow managed to get onto a cargo ship. Their descendants now cover the island at a density of about 500 snakes per hectare. These snakes, which grow to more than 2 meters in length, are able to climb trees and feed on young birds in the nests. The birds of Guam, like those of other oceanic islands, have evolved without such predators and so have no evolved defense.

Adaptations of the Brown Tree Snake include very large eyes (for finding prey during the day) and a remarkable capacity to move rapidly up and down trees as well as along the ground. This "introduced" predator on Guam has caused the extinction of many bird species and is a major threat to the survival of others. Biologists expect the Brown Tree Snake to arrive on other islands, as stowaways on cargo ships, and are particularly concerned about Serenity Island, where bird species are already endangered by a variety of introduced enemies.

The field biologists have provided you with a life table (Table 22.1) and the following information.

The major cause of mortality during the first year of life is humans, who have learned to eat the snake eggs. In the second year of life, the young snakes disperse and are preyed upon by shrikes (a sit-and-wait predatory bird) as they move through the forest in search of a less crowded place to feed and to reproduce. The shrikes would take more young snakes if they could see them in the dense vegetation. When the snakes are 2 years old, they are larger and are preyed upon by herons (a sit-and-wait predatory bird), which also

Table 22.1 Life table for the Brown Tree Snake

x	l_x	p_x*	m_x	$l_x m_x$	F_x**	*Major cause of mortality* (for descriptions of models and variables, see the relevant exercises in this text)
0	1		0			humans
1	0.8		1			shrikes ($a = 0.1$; $b = 0.003$; $d = 0.07$)
2	0.24		1			herons ($a = 0.08$; $b = 0.004$; $d = 0.06$)
3	0.144		2			humans
4	0.1152		3			humans
5	0.0922		5			humans
6	0.0737		6			humans
7	0.0590	0	6			old age

*$p_x = l_{x+1}/l_x$
**$F_x = p_{x-1} m_x$

would take more snakes if they could see them in the dense vegetation. After that their main enemy is humans, who kill snakes for food and out of fear. You are surprised to find that snakes reproduce a year earlier and have much larger clutches than populations in California. The embryos develop inside eggs (like in a bird), which are deposited in burrows about 30 cm below the soil surface. When the eggs hatch, the young are well developed and able to feed themselves. The very high rate of reproduction and superb ability to disperse have resulted in a wave of invasion that is moving outward from the port on Serenity Island.

The Starling

The Starling (*Sturnus vulgaris*) was imported from Europe to North America in 1890 by a New Yorker who decided that Central Park in his home town should have all the bird species mentioned in Shakespeare's works. The Starling is an aggressive omnivore that nests in tree holes. It quickly multiplied and populations spread outward, away from New York City. The distribution of the Starling now extends from the Atlantic to the Pacific and from the Mexican border to the Canadian border. One of the many problems resulting from the introduction of the Starling is the near extinction of the Eastern Bluebird (*Sialia sialis*). The two species compete for nesting sites, in tree holes, and the more aggressive Starling always wins. The advantage of nesting in tree holes is protection of the young from nest predators.

Table 22.2 Life table for the Starling

x	l_x	p_x*	m_x	$l_x m_x$	F_x**	*Major cause of mortality* (for descriptions of models and variables, see the relevant exercises in this text)
0	1		0			predation by pine martens ($a = 0.2$; $d = 0.05$; $b = 0.001$)
1	0.9		0.5			predation by hawks ($a = 0.3$; $d = 0.06$; $b = 0.001$)
2	0.45		1			aspergillosis*** ($b = 0.004$; $p = 0.02$; $u = 0.001$; $d = 0.05$)
3	0.315		2			intestinal worms ($b = 0.03$; $d = 0.01$; $a = 0.05$; $c = 0.01$; $u = 0.01$; $p = 0.03$; $m = 0.01$)
4	0.2205		2			intestinal worms ($b = 0.05$; $d = 0.01$; $a = 0.01$; $c = 0.005$; $u = 0.01$; $p = 0.02$; $m = 0.01$)
5	0.1544	0	4			old age

*$p_x = l_{x+1}/l_x$
**$F_x = p_{x-1}m_x$
***a fungal growth on linings of the respiratory tract (fungi are a form of microparasite)

The field biologists have provided you with a life table (Table 22.2) and the following information.

From the time the eggs are laid until the fledglings leave the nest, the young are quite safe within the tree. Pine martens occasionally prey on young Starlings, but do not have a major impact on the survival rate of that age group ($p_x = 0.9$), because they specialize on squirrels. Between ages 1 and 2, the young Starlings disperse in search of their own place to live. Those that find a place begin to reproduce immediately. The clutch is small (two eggs) this first year. During this time, the inexperienced 1-year-olds are vulnerable to predation by hawks. While there are lots of juvenile Starlings to eat, the hawk populations are not accustomed to hunting them. From years 2 through 5, the clutch size increases steadily and some of the adults suffer first from a fungal growth and later from intestinal worms (which increase in number as the bird grows older). As with other "introduced" species, the Starling does not have a high worm burden because the worms adapted to the Starling are still in Europe. Finally, at the end of the fifth year, most of the Starlings die of old age. Compared with other birds, the Starling has a high rate of reproduction and a short lifespan.

The Gray Squirrel

The Gray Squirrel (*Sciurus carolinensis*) is native to the deciduous forests of eastern North America. It was deliberately introduced (because it is cute) into England in

Table 22.3 Life table for the Gray Squirrel

x	l_x	p_x*	m_x	$l_x m_x$	F_x**	*Major cause of mortality* (for descriptions of models and variables, see the relevant exercises in this text)
0	1		0			predation by stoats (a form of weasel) ($a = 0.001$; $b = 0.01$; $d = 0.08$)
1	0.7		0.5			death on roads during dispersal ($a = 0.005$; $b = 0.01$; $d = 0.01$)
2	0.56		1			predation by domestic cats
3	0.448		2			coccidia (microparasite) ($b = 0.008$; $p = 0.01$; $u = 0.01$; $d = 0.07$)
4	0.224		3			intestinal worms ($a = 0.03$; $c = 0.1$; $u = 0.01$; $p = 0.2$; $m = 0.1$)
5	0.112		3			liver flukes ($a = 0.05$; $c = 0.3$; $u = 0.01$; $p = 0.1$; $m = 0.005$)
6	0.056	0	4			old age

*$p_x = l_{x+1}/l_x$
**$F_x = p_{x-1}m_x$

1876. From a few individuals in London's Hyde Park, the Gray Squirrel population has swelled to 2.5 million and fills the deciduous forests throughout the island of Great Britain.

Two problems have arisen from the presence of the Gray Squirrel in England: (1) damage to the forests from the squirrels feeding on phloem (which they rarely do in North America) and (2) decline of the Red Squirrel (*Sciurus vulgaris*), which the Gray Squirrel replaces wherever they come into contact. A population of Gray Squirrels has now appeared in Norway, where your field workers have compiled a life table (Table 22.3) and the following information.

Predation by stoats is low, because they are not accustomed to hunting the Gray Squirrel. Domestic cats take some squirrels, mainly those between 2 and 3 years old, because at that age they are too young to compete for nests high in the tree, where the cats are less likely to climb. Moreover, the cats are too well fed to be a significant mortality factor. You are surprised to see that the mortality from dispersal is also low compared with that in the United States. Microparasites and macroparasites also have a low impact relative to their effects on squirrels in North America, because the squirrels lost their parasites during the process of colonization.

Table 22.4 Life table for the Asian Longhorned Beetle

x	l_x	p_x*	m_x	$l_x m_x$	F_x**	*Major cause of mortality* (for descriptions of models and variables, see the relevant exercises in this text)
0 (eggs hatch)	1		0			bad weather
1 week (0.5–1 mm)	0.9		0			fungal disease (microparasite) ($b = 0.02$; $p = 0.02$; $u = 0.5$; $d = 0.05$)
2 weeks (1 to 2 mm)	0.72		0			fungal disease (microparasite) ($b = 0.04$; $p = 0.01$; $u = 0.2$; $d = 0.05$)
3 weeks (2–3 mm)	0.648		0			nuthatches ($a = 0.001$; $b = 0.002$; $d = 0.1$)
4 weeks (pupa)	0.4536		0			woodpeckers ($a = 0.001$; $b = 0.01$; $d = 0.1$)
5 weeks (flying adults)	0.2268		0			barn swallows ($a = 0.005$; $b = 0.01$; $d = 0.1$)
6 weeks (adult lays 200 eggs in a new tree)	0.1814	0	100			genetically programmed death after reproduction

*$p_x = l_{x+1}/l_x$
**$F_x = p_{x-1} m_x$

The Asian Longhorned Beetle

The Asian Longhorned Beetle (*Anoplophora glabripennis*), which is native to the maple forests of northern China, was accidently brought into the United States (in maple wood) in the 1980s. From a few individuals in New York City, the population spread northward. The beetle lives within the woody parts of maple trees where it damages the phloem, blocking passages that normally carry sugar and other nutrients from one part of the tree to another. An infected tree dies within a few years. The Asian Longhorned Beetle has just appeared in southern Vermont, and the tourist industry in that state has asked to have the spread of this pest brought under control.

You hire a team of field biologists, who study the life history of the beetle where it occurs in Massachusetts. They provide you with a life table (Table 22.4), which gives a 7 by 7 Leslie matrix, and the following information.

This beetle population has a high rate of growth (per week) because it was "introduced" without its natural enemies – parasitoids in China. In late summer, the adult female lays 200 eggs (half are females), which overwinter in a dormant state. In May, the eggs hatch. Between the time that a beetle hatches until it is 3 weeks old, the larva is vulnerable to attack by a parasitic fungus that lives within the wood of the tree. After that, as the larva grows and feeds on the tree, its main cause of mortality is predation by nuthatches. When it is 4 weeks old, the larva becomes a pupa and its main source of mortality is woodpeckers. After a week of transformation, the beetle emerges as an adult. The adult form, with wings, leaves the maple tree and flies about in search of a mate. At this time, the adult is vulnerable to predation by barn swallows. After mating, the male dies and the female searches for a new tree to invade. Once she finds a suitable maple, she bores into the wood, lays her fertilized eggs, and then dies.

Glossary of MATLAB Commands

abs find the absolute value

dn differentiate *n* with regard to time

dsolve find the symbolic solution of the ordinary differential equation

eig give eigenvalues and eigenvectors

eval interpret the text containing MATLAB expressions

exp base of the natural logarithm

figure set up a graph

function a form of program ("m-file") that can accept input arguments and return output arguments; internal variables are local to the function

global declare the variables as parameters to be passed to the differential equation

hold on keep the graph available for more plotting

max find the maximum solution to the equation

ndot(1,1) arranges for the function to return the column vector dN_1/dt

ndot(2,1) arranges for the function to return the column vector dN_2/dt

nhat the number of individuals at equilibrium

nprime predicted number of individuals using equations for discrete time

ode23 use ordinary differential equation solver # 23 to find the numerical solution to a differential equation

ode45 use ordinary differential equation solver # 45 to find the numerical solution to a differential equation (`ode45` uses larger steps and returns fewer data points than `ode23`)

plot present the mathematical relationship on a graph

Population Ecology: An Introduction to Computer Simulations. By Ruth Bernstein.

pretty present the formula in a more conventional style

rand('seed') draw numbers from a random number generator

simple find the simplest form of the equation

simplify simplify the symbols

solve find the symbolic solution of the equation

subplot prepare a graph that is part of a larger graph, formed from more than one graph

subs substitute one symbol for another

sum sum the elements

sym create, access, or modify a symbolic matrix

symop perform a symbolic operation

tspan duration of the simulation

vectorize make a version of the equation that can be plotted on a graph

+ add

- subtract

* multiply

/ divide

^ exponent

( ) solve this part of the equation first

Index

Color, Marker, and Line Style for the Graphs

If you do not specify a color, MATLAB presents the first curve in blue and then, for additional curves on a graph, moves sequentially through the seven colors as listed in the table below. If you do not specify a line marker, no line marker is drawn. If you do not specify a line style, MATLAB uses a solid line. You can specify your colors, markers, and line styles by entering one of the following symbols at the end of a plot command.

Symbol	Color	Symbol	Marker	Symbol	Line style
b	blue		point	-	solid line
g	green	o	circle	:	dotted line
r	red	x	x-mark	-.	dash-dot line
c	cyan	+	plus	--	dashed line
m	magenta	*	star		
y	yellow	s	square		
k	black	d	diamond		
w	white	^	triangle (up)		
		<	triangle (left)		
		>	triangle (right)		
		p	pentagram		
		h	hexagram		

Source: *The Student Edition of MATLAB*, Version 5: User's Guide, 1997, page 187. Reproduced with permission from The Mathworks, Inc.

Population Ecology: An Introduction to Computer Simulations. By Ruth Bernstein.